NORTH
AMERICA

SOUTH AMERICA

kansas city public library
kansas city, missouri

Books will be issued only
on presentation of library card.
Please report lost cards and
change of residence promptly.
Card holders are responsible for
all books, records, films, pictures
or other library materials
checked out on their cards.

photo: Howard Johnson

ALONE AT SEA

Hannes Lindemann
EDITED BY JOZEFA STUART

© Copyright, 1958, by Hannes Lindemann
All rights reserved under International and Pan-American Copyright Conventions. Published in New York by Random House, Inc. and simultaneously in Toronto, Canada, by Random House of Canada, Limited.
Library of Congress Catalog Card Number: 58-9874
Manufactured in the United States of America by H. Wolff, New York

CONTENTS

First Voyage

1 THE START OF MY VOYAGE, 3
2 WHITE SHADOWS IN THE GULF OF GUINEA, 10
3 EMERGENCY LANDING, 28
4 THE BIG JUMP, 50
5 THE LAST STORM, 85

Second Voyage

6 RESOLUTIONS AND PREPARATIONS, 103
7 AN IMPOSSIBLE VOYAGE, 109
8 CONCLUSIONS, 177

Voyage

1 THE START OF MY VOYAGE

From a very early age I have loved the sea and sailing. When I was a small boy, my grandfather, a sailor from the old windjammer days, stirred my imagination with the lore of sailing and the legends of the sea. Under his guidance I first learned how to handle a boat. As I grew up, my interests and activities widened to include long trips in sailboats and in folding boats—small, kayak-like boats with collapsible wooden frames and rubberized canvas covers. I sailed the rivers of Europe, and when my skill and confidence increased, I sought more exciting voyages. Single-handed, I rounded the Iberian Peninsula and sailed through the Mediterranean. Out of these experiences gradually grew the idea for the greatest adventure in single-handed sailing—an Atlantic crossing.

In all of us there is an impulse—though it may be deeply hidden—to leave behind us our ordinary lives and go beyond the morning to seek our fortunes. This urge is usually thwarted in our time by the restricting responsibilities of family or society. Yet some continue to climb almost inaccessible mountains or to explore the distances of the sea, dreaming of other coasts. And the curious thing is that when this impulse comes to the fore in some individual and is acted on, most men are puzzled; so remote and fantastic, perhaps, do their own dreams seem.

I am a doctor by profession, trained in Hamburg, where I always intended to settle down and practice. But restlessness and curiosity drove me instead to travel and work abroad. In 1952, when I was twenty-nine, I found a job at a U.S. air base in French Morocco, and while there I signed a two-year contract to work in the plantation clinics of the Firestone Rubber Company in Liberia.

When I was working in Morocco I had met a man who, as a voluntary castaway, had studied the problems of survival at sea. One of his most firmly held convictions (which came to be widely known) was that it is possible for a castaway to survive by drinking salt water. I found it impossible to accept his thesis. I was convinced that acceptance of such advice might easily endanger the life of a castaway, that the human body is not capable of surviving the rigors of exposure and the danger of dehydration without recourse to fresh water. I felt challenged both as a doctor and as a sailor to put his theory to the test myself.

The idea of experimenting with the problem of survival at sea continued to excite me after I moved to Liberia. In my free time, while tropical downpours drummed on the roof of my bungalow, I studied books on boats, sailing and the experiences of other single-handed sailors. By the end of my first year in Liberia, I decided the time had come to plan seriously for an Atlantic crossing. My first step was to acquire a boat. Clearly, I could have done what so many have done before and bought a small sailboat, but, living in Liberia, where the dugout canoe is the vessel of all native fishermen, I was inspired to try one of them. This would be original and exciting: to sail across the ocean in the most primitive of all boats. If, as some scientists believe, an early cultural tie may have been established between the West African coast and the Caribbean Indians by early canoe voyages, I would be emulating the explorers of prehistory. In any case, to test my survival threshold and my seamanship, I would remove myself as much as possible from the crutches of our comfortable civilization.

I had the choice, when I first started making my plans, of buying a second-hand canoe or of building one myself. As I had twelve months in which to make my preparations—I planned to leave Liberia as soon as my Firestone contract expired—I decided to build one. In that way I could be certain of the strength of my canoe, which would have to withstand battering Atlantic waves. Also the boat had to be carefully designed in weight and balance to be able to ride out storms without capsizing. I knew I would have to make certain modifications in the crude coastal canoe of the West African fisherman. So I decided to begin at the very beginning and pick out a tree in the jungle that I could fashion into a suitable dugout.

For its strength and size, and because I knew that the Fanti fishermen of Ghana use it, I chose a kapok, a common West African tree, which can grow to a height of one hundred and eighty feet and a diameter of six to nine feet. Without much difficulty, I found a tree suited to my purposes, growing on the territory of one of the local paramount chiefs. I explained my need for the tree to the chief and offered to buy it from him, but he insisted I take it as a gift.

My troubles began after I had the tree. I started enthusiastically and innocently, unaware of the difficulties that any unusual venture in the tropics is sure to encounter. I offered the job of cutting down my tree to three stalwart young men. After studying the tree, they refused. So towering a giant, they claimed, must be the home of evil spirits, who would revenge themselves for the loss of their tree by taking a human life. I offered more money, but their fear of the spirits was greater than their love for money. I was almost prepared to fell the tree myself when I learned of a village whose inhabitants are professional woodcutters and whose evil spirits do not haunt treetops. Further negotiations with their chief bought their services; one week later, my tree was felled and a thirty-six-foot length cut from the trunk.

I had chosen my tree well; the wood proved healthy and easy to work with. In eight weeks, two young natives, working with axes, chopped out the interior. The trickiest part of hollowing out a canoe comes when one tries to get an even thickness of the trunk walls. Our method was crude and simple: we chopped on the inside with a transversal axe and held our hands to the outside to get a sense of the thickness of the trunk. At the end of eight weeks, we carried my roughly hewn boat to my house on the plantation and stored it under the porch. Once a week I sprayed it with insecticide (a necessary precaution in a tropical climate), meanwhile looking for a skilled carpenter who could finish the job. I found Alfred. His first contribution was to write on the stern: This boat is sixty-four feet long. My two houseboys were very much impressed by Alfred's erudition; I less so, for the boat measured only thirty-six feet. Alfred's carpentry proved no better than his mathematics; so I looked around for a replacement. My next helper was William More; but, as it turned out, he could not work unless he got his daily ration of fermented cane juice. And sometimes he could not work when he did. Despairing of reliable carpenters, I set to work to do the job myself with the help of my two houseboys.

After four months of hard work the canoe was finished, except for the keel. We drew the boat up in front of the house and set to work smoothing the final rough spots. To my consternation our planes uncovered insect holes. Out of my boat crawled fat white maggots, small black bugs, big black wood beetles and bark-colored stag beetles with antennae as long as my finger. Lying for eight weeks in the jungle, the trunk had become a haven for the rich, varied insect life of the rain forest. The insecticide, which I had sprayed and rubbed on the wood with such care, had betrayed me. Hoping that I might be able to smoke out the insects, I asked Sunday, my houseboy, to light a smudge fire under the canoe. The biting smoke forced me away from the house. I returned a few hours later to find the *Liberia*—as I had christened the canoe—and six months' hard work, burning brightly. Sunday slept peacefully beside the bonfire.

I started afresh the next day on my search for a canoe. I was still hopeful of acquiring a new one; so I visited a canoe-building tribe in the interior. I made them the tempting offer of four times their usual price, and they promised to do the job for me. My contract with Firestone had only another six months to run; time was therefore precious to me. But it held no meaning for them; despite my urgings they did not begin the work. I now realized that I no longer had time to build my boat, that I would have to make do with a second-hand canoe. I found one, belonging to a fisherman of the Fanti tribe, which seemed suitable; I offered twice the price of a new canoe for it, only to be disappointed again; the fisherman, who had at first been willing to sell his boat, changed his mind at the last minute.

I had a friend among the local fishermen, a Liberian named Jules. Now I went to him, in desperation, begging him to help me find a boat—no matter what the quality. One week later he found one for me. It had holes in the stern and bow, and in the bottom where it had lain on the ground. Also fungus growth had softened the wood somewhat. Still, the trunk seemed strong enough, and in any case I planned to strengthen it further by covering the hull with fiberglas.

Now, finally, three months before my hoped-for departure, I at least had a boat. The mahogany canoe measured twenty-three and a half feet from bow to stern on the outside and twenty-three feet on the inside. Its width was twenty-nine and nine-tenths inches outside and twenty-eight and seven-tenths inches inside. My houseboys and I made a keel five-and-one-tenth-inches deep and eleven-and-a-half-feet long and weighted it with two hundred and fifty pounds of lead. We planed the underside of the trunk with an electric sander and painted it with a mixture of hardener and resin. Using this mixture as an adhesive, we attached fiberglas to it, and then painted it over several times with the same mixture. This process was necessary to ensure the strength of the trunk.

When the hardener and resin were thoroughly dry, we set the boat on her keel and spanned her width with bent lengths of iron. We made a deck by covering them with plywood, leaving a small cockpit in the stern and a hatch before the mast. At the approximate water line on either side of the hull, we attached corkwood pads—each some ten inches thick—hoping they might lessen the roll of the boat. The canoe now resembled the pirogues of the Carib Indians. We also covered the deck with fiberglas. It afforded additional strength and would also protect the wood against teredos, a shipworm or destructive mollusk prevalent in West Africa.

Bearing in mind the possibility that I might capsize, I took measures to ensure that the canoe stayed afloat by partitioning off the ends with bulkheads, putting empty airtight containers behind them. I attached steering cables to the rudder so that I could control it with either my hands or feet. Then we made a mast of ironwood, which has enough give to it so that the boat could run even in the Gulf of Guinea without a backstay. The boom was made from rare red camwood, which warps even less than mahogany.

The long-awaited day of launching arrived exactly four weeks before my contract expired. Slowly and carefully, we drove the boat on a company truck to Cape Palmas, eighteen miles away. I stood, movie camera in hand, while my friends did the launching. But the canoe would not stay afloat; the shallow keel was too light to counterbalance the weight of mast and sail. We filled three big sacks—provided for this eventuality—with sand and used them as weights. They gave us sufficient stability and, with a three-horse-power motor, my canoe made her first test run. In memory of my ill-fated first canoe, I christened her the *Liberia II*.

I accustomed myself to the handling of the canoe by making short sailing and fishing trips. During these I found that a jib of three square yards, a square sail and a gaff sail of nine square yards gave me sufficient play in varying winds.

I registered the *Liberia* as the first "yacht" in Cape Palmas and loaded her with a three-month food supply. With Haiti, first Negro republic of the world, as my destination, I set sail one hot February day.

2 WHITE SHADOWS IN THE GULF OF GUINEA

I left the little harbor town of Harper, which lies in the lee of Cape Palmas, with two young boys—paddles in hand—perched on the bow of the *Liberia II*, while I sat in the cockpit. Jules paddled alongside in his canoe. He was to take the boys back after we had got my boat out of the harbor. At the sight of my unorthodox craft even the fishermen and dockside loungers were startled out of their usual apathy. Their interest increased when, just as we were making headway on the outgoing tide, cries of "Stop, Doctor," forced me to look back. Customs officers were signaling to me to return. I shouted across the water that my sailing papers were in order, as indeed they were; I had paid my taxes, tipped in the right quarter, and I had no intention now of postponing my departure.

We continued out to sea, to the southwest, paddling past a steep rock that pointed an accusing finger into the open ocean. I noticed three friends waving from the roof of a building on the rock, and I answered by clasping my hands over my head. To the left, high up on the cape, we passed a large building. People breakfasting on the roof garden there looked down on us through binoculars and stretched out their hands in a well-bred, unenthusiastic farewell wave. I was irritated by their superior attitude, which seemed to me to imply a complete lack of faith in my chances of success.

The *Liberia* now faced her first real test of seaworthiness. We had to cross a reef where waves reflected from the rocky coast met the large wind-blown waves of the ocean. Slowly we struggled through a seething, foaming mass of shallow water. The boys on the bow paddled hard, and we made it. I cast a last look at the fishermen's huts that lined the harbor. The water deepened, the waves flattened, and I took deep breaths of the pure air of the open sea, happy to leave behind the typical harbor smell of rotting fish and decomposing rubbish mixed with the salt of surf spray.

But we were still not completely free of the harbor; a channel, lying between the rocky cape and a small island, had to be negotiated. Ocean waves and waves reflected from the shore mingled there, and the tide from the strong Guinea Current struggled around the entrance. Once again, the two boys paddled me through.

The time had now come for me to go on alone. The boys jumped into Jules's boat, I hoisted my after sail and Jules shouted, "We will pray for you, Doctor." I was deeply touched by his farewell, and even forgave him his lapse of the night before. I had asked him to watch my boat for me while I went out for a few last drinks with my friends. On my return I found he had let her capsize in the outgoing tide.

The outboard motor was waterlogged and refused to come to life again. Jules's carelessness forced me to leave without a motor.

As I sailed with the current in a southeasterly direction, I sat on the windward side of the boat and gazed back at the sandy beach of Harper. Gray spray hung like a silk curtain between us. After two years in Liberia I had formed a strong attachment to the country and its people. It was not so much the work I did, though it was more responsible and freer than any I had known before, but the warm-hearted, generous people that I knew I would miss. I had chosen Haiti as my destination largely because I was eager to see if the only other Negro republic in the world had the same unspoiled spirit.

Gradually I lost sight of Harper, and some miles east of Cape Palmas I saw the coconut palms which shaded the round huts of the small village of Half Grevy. Columns of smoke from native bush fires drifted lazily to the sky. Hanno of Carthage, who sailed down the West African coast in the fifth century B.C., has left us an account of these same bush fires. There has been little change in the way of living here from his day to ours; the men burn down the bush for new land, women clear the land, hoe it and sow hill rice between logs. After each harvest the land lies fallow for two or three years, and during that time other parts of the bush are burned off.

As the sun stood at its zenith, I passed the southernmost cape of this part of Africa. The rising tide at the mouth of the Cavally River pulled the boat shoreward. The tremendous force of the Cavally as it spews its waters into the ocean changes the shoreline here almost every day. When the tide falls, dirty inland waters mingle with the raging surf over the sandy reef and attract an immense number of barely visible inhabitants of the sea and their bigger brethren. I heard the seething thunder and roar of waves and saw hungry, foaming breakers run rhythmically up the beach and lick the yellow sand. Here the swell, eternal breath of the ocean, has ground the stones of the beach to powder—as though, like Sisyphus, damned eternally to push stones up and down, up and down the beach. I gazed at the inhospitable West African coast from the canoe, seeing what has met the eye of every seafarer who ever traveled here: an unending stretch of waves hurling themselves at a flat beach, unbroken by harbors or protected inlets that can offer refuge to the sailor. The sole signs of human life are the bush fires of the Africans, and only an occasional mangrove swamp or huge gray boulder interrupts the monotony of sea and sand.

It is not an easy coast to sail along, and I was thankful when the time came to leave it. To the north the white beam of the lighthouse in Tabu grew smaller and smaller. A school of minute octopi leaped out of the water with the force of jet propulsion and then—sk-latsh—fell back into the sea. The sun sank behind a chain of clouds with a rare and exciting display of color. It was as if this short, dazzling eruption were to compensate for the monotony of the dark night ahead. My horizon was bounded briefly by a bank of golden clouds and then, with startling suddenness, the bright tropical bloom darkened to a threatening blue-black. At the beginning of my first night at sea, I felt as though I were part of an overwhelming natural spectacle.

The first stars shone high above now; shadowy waves raced past the dugout, and occasional small combers lapped against the hull. It was time for me to be practical and think about my first meal. Can a castaway survive by drinking sea water? I intended to find out and, therefore, planned to drink one pint every day. I knew that any amount beyond that would permanently damage my kidneys. I carried canned milk and fruit juices with me because I knew I would also have to drink other liquids, or my kidneys would be unable to excrete the high percentage of salt. (I never did accustom myself to the taste of sea water and at each first swallow I came close to vomiting. Thereafter the salty taste would disappear because of an augmented secretion of saliva.) I had not eaten in the twelve hours since leaving Harper, and I was surprised to notice that I was not hungry. My energy level had already lessened and it took me a long time to make decisions. Despite my lack of appetite, I knew I had to eat, and out of my supplies I chose sausage, butter and black bread for my first meal. I managed to choke down two slices of bread, and as I ate I thought back to the days of the old, slow sailing ships, when food for the crew was an even greater problem than navigation.

My meal was interrupted by a sudden burst, an explosion, beside me. A ray hit the water with its hollow wings. I sat quietly but heard no more from him.

The wind weakened. I fixed the tiller, squeezed my body underneath the plywood deck and put on my wind jacket. Meanwhile, the pirogue had sailed through the wind and gone off course, and I had to bring her back with my paddle. Glancing overboard, I noticed with surprise that the bow wake water shone in silver-gray streaks. Even the combers glowed. I hit at the surface with my paddle and produced a new kind of fireworks display: out of the water rose phosphorescent particles, big and small, which shone yellow, orange, green and blue in the darkness. Dense clouds of bioluminescent plankton had risen to the surface; its movement created this nighttime magic.

While I amused myself by slapping the water with my paddle, I noticed the lights of a ship far away. As I sailed closer to it, I took out my flashlight and beamed it onto my gaff sail, hoping to make my presence noted. The ship appeared to be stationary; then suddenly it veered off in a southeasterly direction. Time passed slowly. I sat, tiller in hand, with nothing to do, struggling to stay awake. Because I was on a steamship route and there was a danger of collision, I had decided to sleep during the day and sail at night. In daylight I could be seen, but at night I would have to signal. Even so, I was not now sure that the beam of my flashlight was enough to warn other ships of my presence.

The morning came at last, bringing with it a visit from a shark. It was a nine-foot-long tiger shark, and I was curious to see what it would do. Cautiously circling the boat, it gazed up at me with ugly little pig eyes; it thrashed its dark brown, spotted tail, and drops of water landed on my deck. Several little pilot fish swam nervously back and forth between the rudder and their master the shark, which continued on its way, undisturbed by its excited retinue. After a few rounds of my boat, it suddenly disappeared; I must confess I felt a certain relief.

A calm breeze wafted over the sea now, enough to fill the sail and cool my sweating body. Throughout the morning, shining mackerel jumped out of the water around me. At noon I took my position with my bubble sextant, and found I was nearly sixty miles south of Cape Palmas. At this time of day the intense heat of the sun beat down on the unprotected crew of the *Liberia* with tropical intensity. I draped a wet towel over my shoulders to prevent a serious burn on my neck; I dangled my feet over the edge; finally I decided to try a cooling bath. I slid into the water and swam around the dugout. Unfortunately, I had not brought underwater goggles with me, so I missed the tremendous variety of life that abounds in the ocean. My bath did not refresh me as much as I had hoped it would. Afterward I sat in great discomfort while the sweat and salt water ran off my body, mingled and dripped into the bilge.

All day long I had enjoyed watching the petrels as they fluttered in the swell, pecking at plankton, or held their wings high, seeming to walk on the surface of the ocean with their webbed feet. Sailors of long ago christened them petrels after St. Peter; like him they seem to want to walk on water. (I was grateful to all the living creatures of the sea and air that I could watch from my boat. They shared my solitude and were a steadying reminder of everyday life throughout my voyage. They helped to keep me from total absorption in my daydreams, and later they even dispelled desperation.) As I watched the starling-sized, soot-black petrels I was reminded of the agitated flight of bats. Petrels are known as birds of the night, but I cannot imagine them more active than they were during the day. In the daytime I never saw them sleep and I rarely saw them alight on the water for longer than a minute.

A blood-red sunset ushered in my second night at sea. In the twilight a petrel flew carelessly into my gaff sail but escaped unharmed. The night passed uneventfully. For an active young man the sea was too smooth and too calm; the hours dragged slowly by. I took short cat naps, amused myself with dreams of my future, and sang songs from my college days.

When the third day dawned there was still no wind. I fell asleep and awoke in the heat of noonday, bathed in sweat. I took my position and to my disappointment found I had made only half as much progress as on the day before. My feet were beginning to swell from the salt water I had drunk. I was still not hungry. Perhaps it was because I was worried about the boat. I was beginning to realize that there was something radically wrong with the *Liberia*'s ballast. I needed a heavier keel, more outside ballast. The sandbags put in at the launching were not the answer. Still, I was not yet ready to turn around; I wanted to test the boat in stronger winds. I decided to sail as far as the equator, where I could expect to meet the trade winds, to make a final decision. I knew that my return trip, if necessary, would be easier sailing and would take far less time than the outward voyage.

That day I spent my time watching mackerel. They leaped out of the water with such enthusiasm that I was reminded of myself jumping into a cooling bath at the end of a day's work in the Liberian heat. Their bodies were heavy and muscular, and in the air they lacked the stiff elegance of flying fish or the pliancy of a dolphin. They landed on the water with the full weight of their two-foot-long bodies, sometimes flat on the abdomens, sometimes slightly turned to the side.

Toward evening I noticed albacores for the first time. Albacores are large members of the mackerel family; these were three feet long and had long pectoral fins. The surface frothed from their activity, and the metallic sheen of their bodies and the yellow of their tails produced a fantastic display of color in the gold of the setting sun. They held their bodies stiff and straight when they ventured into the air, as though they feared they might shatter like glass.

Just as the sun sank below the horizon I was startled out of my daydreaming by a light thump against the port side of the canoe. A mysteriously large fish surfaced on the starboard side, touched the rudder cable and then submerged without giving me time to identify it. Despite its size I was not very disturbed, as I knew that even a big fish would have trouble capsizing the *Liberia*, and that fish generally yield the right of way. Of course, a whale, surfacing straight from the depths, could lift a small boat out of the water as Joshua Slocum, who sailed around the world in his famous yawl the *Spray*, reported, but it was one of the least of the dangers I faced on my voyage.

The third night passed uneventfully; I searched the skies in vain for the polestar, which I had never seen in West Africa. I was thwarted by a cloudy sky and a hazy horizon. At intervals during the night I heard the childlike, shrill cries of petrels. They sounded doubly loud, for the wind slept, the moon seemed anchored in the sky, and the smooth swell rose and fell, without a wave marring the surface.

In the morning the wind freshened, and splashes of water fell onto my dirty deck. To make the boat sail faster and to hold her upright, I perched far out on the windward side and braced my feet under the cleats on the lee. I sailed swiftly to the southeast until noon, when it was time to take my position. The opportunity to climb down from my uncomfortable seat on the gunwale of the canoe was welcome. Now, as if to facilitate my navigation, the wind died down and the sail swung loose drawing with it the squeaking line. A blue shark, approximately nine feet long, circled lazily twice around the canoe, then swam away, and for a few brief moments the ocean resembled a quiet mountain lake. Then the surface shivered, dark patches appeared in the water, and an unseen hand painted small ripples on the blue canvas of the sea. The sails filled. In the distance I heard a musical cadence of babbling, chattering and subdued giggling; it floated across the water and I imagined I heard the gossip of water nymphs. It soothed me, and my tired, sweating body was revived by the wind. But soon the pleasant, far-off music changed to an awesome rumble that betokened a storm; combers, large and bold, slapped against the hull, coughing and sputtering on all sides. From afar I heard the roar of ever-larger breakers and I was reminded of the rush of coming rain in the tropics. The wind whistled with force through the sails and beat against the mast. My pirogue was too crank to sail athwart; so I took in the sail and put out the sea anchor. (The sea anchor—a framed cone of canvas, which, when in use, is dragged behind the boat, the larger end toward the stern—kept the waves from damaging my canoe and held the stern straight to the wind.) The temperature had dropped with the rising wind, and to protect myself against the cold, I eased myself down under the spray cover until only my head was visible. The *Liberia* rode out her first Atlantic storm successfully, and it was not long before the squall was over, leaving only a gentle, southerly breeze.

By nighttime the weather had changed again. Brooding, menacing cloud banks gathered on the horizon, a screen for a colorless sun that slipped quietly behind them. In the rapidly increasing darkness, these clouds seemed to threaten me personally. Soon sky and clouds merged into stygian blackness, through which the *Liberia* sailed haplessly. As I watched the rise and fall of the bow increase, I found myself wishing I were not alone on a dark night on the Atlantic.

I had no idea of the kind of weather that lay ahead, so I reefed the sail by turning the boom. The threatening weather flattened the swell, and neither lightning nor thunder, wind nor rain relieved the tension. Only streaks of glimmering bioluminescence, whirling like dust in the wake of my pirogue, lightened the heavy, suffocating darkness. Profound silence —then, out of the blackness came a strangely human sigh. The sound rose and fell, softer and louder, like the moan that heralds the approach of death; a voice from an unknown, mysterious source. I could not place the sinister sound, and my ignorance made me afraid. I cursed the unholy darkness through which the pirogue sailed silently, as though drawn by the hands of unseen spirits. As we approached the lament, it ceased, and no beat of waves nor fluttering of bird wings broke the silence. I shone my flashlight over the water, but I saw nothing. (It was only much later that I read a description of the moaning of petrels during their breeding season and realized that that was the sound that had frightened me so.)

Gradually the silence around me was broken, first by the moderate whispering of the waves and then by a louder and more distinct gurgling. From afar came the rush of wind, the sail filled and the boat listed heavily. To balance the canoe, I had to leave my shelter and sit on the windward side. I tried to make myself comfortable by putting a cushion under me, but it slipped and was lost in the water. I had stupidly forgotten to tie it. I realized that many of my good intentions had already gone overboard in like fashion. I had meant to keep a detailed logbook and to study carefully the reactions of my body to the ordeal. But what, in fact, had I done? Fallen into reveries of my past and plans for my future, or spent hours gazing dreamily at the sky, the water, the fish and the birds. My primitive environment, my low-calory diet and the continued lack of sleep: all contributed to inhibiting my activity.

Now, I sat in discomfort on the side of the boat. The wind rose, tearing up sinister masses of clouds as it raced across the skies. Soon it swept the heavens clean, and the water shimmered with the reflection of stars. Around me, bioluminescent foam glowed like the last embers of a fire, and looking upward, I saw the polestar for the first time, almost engulfed by the high waves but rising again and again.

My next day dawned on a heavy sea. I held to a southeast course, with a 'swell so high that I knew it had just left a up as high as possible for protection. Every hour or so I let the *Liberia* sail a little away from the wind and bailed water out of the bilge.

Slowly the pirogue approached the equator, the area of the southeast trade winds.

On the eighth day the wind strengthened and, changing to the southeast, drove small white typical trade-wind clouds northwestward toward Liberia. I was beginning to need sleep badly, but in such dangerous seas I could not relax my vigilance; so I took pills to counteract my extreme fatigue. Despite the pills and despite the fact that I was precariously balanced on the washboard, I dozed off. Several times I awoke just in time to grab the cleat as I was slipping off my seat. Then I hit on the idea of fastening a line around my body and attaching it to the boat.

At noon the following day I reached a point a few miles south of the equator. The wind still blew with twenty-mile-an-hour force. For three days I had been uninterruptedly on the alert; only cat naps of a few minutes had broken my constant watchfulness. I forced myself to a superhuman effort to keep awake; I sang, I shouted, I screamed at the wind at the top of my lungs, and still I dozed off, to be awakened by knocking my head against the hull. I threw the sea anchor overboard and took in the sails. Then I settled into my cockpit and tried to sleep. Because of my overwrought condition, the combers beating against the hull sounded like thunderclaps. The motion of the boat did not disturb me, but my sleeping arrangements were uncomfortable: my shoulders took up almost the width of the hull, and I had to sit diagonally to be able to move at all. I waited for sleep to overtake me, but I found that my nerves were too much on edge. My legs itched, my back needed scratching, it seemed as though an army of ants marched up my arm;

I lay there, bathed in sweat, in the rays of the merciless tropical sun. From Phoenician times to Columbus' day, sailors were convinced that the sun below the equator was strong enough to burn men and ships. This superstition, which held back the circumnavigation of Africa for hundreds of years, seemed about to be realized in my case; I was slowly being scorched by the fierce sun. My head ached and buzzed, my eyes burned, and I could hardly breathe. Sleep in this intense heat was impossible; I climbed out of the cockpit and hoisted the sails. At last the sun went down, the wind sank, but the high swell and my headache remained.

Malaria! Was it possible that I was suffering from one of my recurring attacks of malaria? I rejected this dread possibility, but the thought kept coming back. I threw the sea anchor over again, determined to sleep this time. I left the sail up and lashed the boom. The *Liberia* rolled terribly now, for without the guiding hand of the wind, she would not lie in the right direction to the swell; instead, she bucked like a stubborn mule trying to throw its rider.

I sat down, my head sank limply on my chest and I yielded to utter lassitude and exhaustion. The word malaria haunted me. I knew I ought to rouse myself and take some antimalaria medicine. I was so tired that nothing seemed of any importance or urgency; nothing seemed worth any exertion. But the instinct for self-preservation drove me to make one final effort. Slowly and clumsily I groped for the pills and swallowed them. My experience in the tropics had taught me that malaria, a cunning enemy, overtakes men when they are at their lowest ebb. I remembered now how often I had urged my patients, faced with a fatiguing bush trip, not to forget their antimalaria doses; here I was now, caught myself. Of course, lack of sleep could also account for my condition, but as I drowsed in the cockpit, I was plagued with the image of Alexander the Great, who had died of malaria at my age, thirty-two.

At last I sat up and looked over the edge of the pirogue. Darkness surrounded me, and mist hung everywhere in the air. Under me, the canoe rolled and heaved. I was tired, too tired to solve the riddle of mist at the equator. I knew, however, that I had to stop the terrible roll of the boat. The sail was up and flapped emptily from side to side. Without knowing what I was doing or why I did it, I hung my heavy blanket over the starboard side, and on the portside, a sea bag filled with cans. Then I sank back into the cockpit. I do not know how long I sat there, but I was suddenly roused to half wakefulness by a splashing and shouting in the water. Where was I? Were people swimming near me? Why did they disturb the quiet of the ocean with their vulgar noise? Slowly I drew my tired body up on deck. The night was extraordinary. Gray—everything gray—no sky. I searched the swirling mist for a glimpse of the noisy swimmers. Then I saw them—blurred shapes, trailing white veils, coming at me across the water in a *danse macabre;* I shouted at them, I swore at them, and they danced away. I felt foolish and ashamed. Why had I driven them off with my curses?

For a night and a day I was held tight in the grip of nightmare and hallucination. I recovered with the setting sun and with only a hazy recollection of what had happened, but I realized immediately that I had been active during the time; crates of cans, a copper container with flashlight batteries, my last rubber cushion and other important possessions had disappeared. I had thrown overboard everything that was in my way. It was a cruel awakening. Later, I came to the conclusion that my hallucinations sprang from lack of sleep and that even a short cat nap of ten minutes would have kept me in possession of my senses. As it was, I now faced the fact that I would have to turn back; too many of my possessions had gone overboard, the *Liberia* needed work done on her keel, and I needed rest. I had little chance of reaching Haiti alive if I continued now across the southern Atlantic.

After I had made my decision, I pulled out the map to find the nearest port on the African coast. I chose Takoradi in Ghana, and on the twelfth day I turned the *Liberia* around and started on my homeward sail. The trade winds blew, the dugout sailed to the northeast at a good speed, while I slept through the night.

The return trip was uneventful, and I was grateful to the fish and the birds who helped me to pass the time; dolphins and flying fish jumped and soared through the air, huge porpoises cut the surface of the water in slow motion, and sharks swam in the *Liberia*'s wake. I had time to notice that a dolphin jumps from the water with the easy mobility of a child at play, whereas the flying fish cuts through the air stiff and straight, propelled by fear of pursuing fish. The poor flying fish!—hunted in the water by dolphins and mackerels and in the air by frigate birds. I felt at one with the fish and the sea around as I sailed back; even my hair was now bleached to the color of sea foam by the tropical sun.

One day I was pleased to discover a new fish that danced in the wake of the boat; it was bright green, with a long forklike tail and a dorsal fin that rose more than an inch above the surface. In the sun its colors sparkled emerald, blue and violet, making it an easy prey to its enemies, and only its speed, which was unusual, saved it from death.

I sighted the African coast fifteen days after I had left it. In the fading light of the late afternoon, I saw palm trees and spotted a wrecked ship that lay to the west of an inlet. In the distance loomed a lighthouse. These landmarks told me that I was approaching the Bay of Axim in Ghana, which lies only some forty miles from Takoradi. Then, as the last golden clouds of the evening hung in the sky, a storm gathered; pitch-black walls of cloud overshadowed my pirogue, the wind slapped at the surface of the water. I took in the sails, put out the sea anchor and drew the spray cover over my head. The wind howled and I heard the roar of coming rain, which fell with tropical suddenness on my deck. Like the beating of a thousand drumsticks, huge rain drops crashed on the *Liberia*, thunder exploded over my head and lightning cut across the sky. Despite the storm, the combers, racing ahead of the wind, gave off their bioluminescent shimmer. Secure in my cockpit, I slept through the whole storm.

The wind abated during the night, making it possible for me to sail on. At dawn I started to take in my sea anchor and was pulling in the line when it was suddenly tugged out of my hands. I tried again—one, two, three, four yards came up easily, and then—whoosh—the line shot out again. The canoe listed dangerously; I was afraid I might capsize, so I dropped the line and threw myself against the opposite side of the boat. A ray or manta had attached itself to my sea anchor and trapped me in sight of the coast. I had to make up my mind whether to wait for the fish to detach itself or to cut the line. I was curious to see the creature that had caught me, but my eagerness to proceed far outweighed my curiosity. I cut the line quickly and was on my way, not without regret at missing my captor and photographing it. I have heard many stories of fishermen of the Red Sea and the Canary Islands, whose poorly anchored boats have been dragged out to sea by rays.

With fair winds from the southwest and support from a strong current, I easily circumnavigated Cape Three Points. But it was not until dark that I saw the lights of Takoradi. I cruised west of the harbor during the night, out of the way of incoming steamers; then, in the morning, I began to sail for the harbor entrance. The light breeze barely filled the sails, and I grew more and more impatient at my slow progress. The swell raced with an eerie sigh of death through the hull of a wrecked ship that lay in my path; to the east more than ten steamers waited to load their freight of lumber and bauxite. I came around an easterly pointing breakwater, took in my sails and paddled against the wind and the outgoing tide. I was making so little headway that I gladly accepted the offer of a police boat to tow me in.

It was seventeen days after my departure from Liberia that I jumped ashore in Takoradi. My knees were a little weak and for the first few hours I felt the movement of the sea, but I had no trouble walking. The chief ill effect was the swelling of my legs. Up to the time of the attack of delirium I had drunk a daily ration of four small glasses of sea water, and on the second day at sea my feet had begun to swell. Gradually the swelling extended to my knees. Small broken blood vessels laced the skin surface of my feet, and when I pressed my thumb against the heel of my foot, it left a deep depression. My ankle bones were sunk in swollen flesh; the sensitivity of the nerve ends had diminished. Massaging my legs twice a day for ten or fifteen minutes and daily exercise had not prevented or lessened the swelling; on subsequent trips, when I eliminated salt water from my diet, I found that my legs remained nearly normal in size.

In Takoradi a policeman took me to the immigration office, where I was informed that I would have to find someone to vouch for me if I intended to stay. I was introduced to the harbor doctor, who offered to be my sponsor and allowed me to moor the *Liberia* alongside his boat.

I had no intention of abandoning my plans for a voyage across the Atlantic, but the keel had to be better ballasted or a successful trip was out of the question. I felt less badly about my miscalculation of weight and balance when I remembered that C. H. Voss had made the same mistake when he rebuilt his Indian canoe, the *Tilikum*, for his world cruise in 1901. Because Hamburg had the best facilities for the work that had to be done, I booked passage home for myself and the canoe and left Takoradi within a few days

3 EMERGENCY LANDING

In the Hamburg docks I had the *Liberia* readied for my second attempt to cross the Atlantic. The inner ballast was removed and an equivalent weight in lead placed under the keel. The rudder, which was too small, was replaced by a larger one, and I had a four-inch-wide plank built around the cockpit so that I could sit there in comfort. After a new coat of paint, the boat was ready. I shipped her to Oporto in northern Portugal; from there I planned to sail to Las Palmas in the Canary Islands and make that my port-of-departure for Haiti.

The arrival of my odd-looking boat in Hamburg caused comment and wild conjecture in the papers. One newspaper announced, "A Viking from North Africa awaits good weather to sail across the Atlantic," and then went on to say that the boat had been sailed by an African along the coast of Africa, through the Bay of Biscay and on to Hamburg, without mishap. Another paper reported that I planned to sail first to Norway and from there to America. I ignored the publicity and sat quietly at home until the *Liberia* was ready and shipped to Oporto, when I left to join her. As the train pulled out of Hamburg, I fell into an introspective and melancholy mood; I wished with all my heart that I had never thought of a voyage across the ocean. I tried to analyze the reasons for my mood. Was it this recent contact with my friends, who were all working at steady jobs, or was it the sea's rejection of my first attempt that had brought on my depression? I fell asleep at last, still trying to find the answer.

On my arrival in Oporto, I went at once to the customs house, hoping to clear my boat immediately. But on the basis of earlier experience with Portuguese customs I should have known that my boat would not be released quickly. In true Iberian fashion I wiled away the days in local cafes until the time had come to try again for the release of the *Liberia*. Again I was frustrated by the slow deliberate approach of the officials, who are dedicated to well-thought-out, long-contemplated action. Every day I sat in the cafes and drank coffee in the company of sad little men in dark suits, returning every few hours to the customs house to badger the officials. I discovered that part of the difficulty was their indecisiveness about how the boat should be listed officially. Happily, within a few days the customs chief was struck with the idea of listing the canoe as a "crate of second-hand goods," and the *Liberia* was then released to me.

Losing no time, I drew the boat up on shore in front of town and set to work preparing her for the voyage. I planned on a crossing of about sixty days and I hoped to accomplish it in June and July. I fastened the rudder, rigged mast and sails, stowed away my cans of food. At the end of the day the work was done, and I floated the *Liberia* alongside a barge. As darkness came, I climbed aboard and fell into exhausted sleep.

Early the next morning, I set sail down the Douro River with a light land breeze to help me. I felt a great urgency to outwit time, and therefore ignored the storm warnings of some passing fishermen. I had no time to spare if my trip was to be accomplished before the dangers of hurricane weather made the attempt too foolhardy. So I sailed on, despite the warning, heartily glad to see the mouth of the Douro behind me. My progress was slow; by afternoon the thunder of surf still sounded in my ears, while sinister cloud banks in the west seemed to restrain the sea breeze. To my relief, the wind rose at last and the first small combers left passing foam scars on the surface of the water. I sailed cautiously, under jib alone, as I always do at the beginning of a voyage, until I am sure of my seamanship again.

The Iberian Peninsula is as dangerous to sail along as the West African coastline, and I knew that it was important to keep a good distance between myself and the breaking surf. As a long-time sailor I knew that fewer accidents occur on the high seas than near a coast.

It was the 28th of May, but the weather was still cold; not even two shirts, a pullover and wind jacket were enough to keep me warm.

My route lay between the coast and the sea lane used by large ships, so I felt I could safely take short cat naps during the night. But the intense cold prevented my sleeping in comfort; my teeth chattered, my hands were numb and cramped, a cold wind blew directly in the cockpit. The dugout took on so much water that I had to bail every six hours. My spray cover was too short to protect me adequately; in fact, it forced the wind directly into rather than over my cockpit. I had made little holes in the hull, through which I ran cables that enabled me to handle the tiller with my feet as well as my hands; whenever a big wave came along, it splashed water into the boat through these holes. It wasn't much compared to the amount that later storms were to force into the *Liberia*, but it was enough to add to my discomfort.

The second evening out of Oporto, the Portuguese north wind, a well-known wind in that part of the world, strengthened to over fifteen miles an hour, and I crawled into the cockpit to sleep, leaving the canoe in the hands of the sea anchor. I had purposely not brought a mattress to sleep on because a hard surface under me would make me, even in sleep, more susceptible to any change in motion. I lay on two crates of oranges and apples, and with even the slightest change in the roll of the boat, I was immediately awake. (I think this saved my life later in the voyage.) After my uncomfortable night on the two crates I was always happy to see the beginning of the dawn, and I looked forward all night to the moment when I could hoist the sails.

I awoke on the third morning to a breeze of some fifteen miles an hour. For a big boat this wind strength is only moderate, but for me it was ideal. With it the *Liberia* was able to achieve her maximum safe speed. On awakening, I collected my rations for the day, before hoisting the sail. Once under way, I could only get at them by letting go the tiller, which invariably put me off course. To reach my food I had to crawl headfirst into the stern, and soon my elbows and knees were covered with sores from the roll of the boat. Yet I had done my best, before my departure, to organize the *Liberia* as efficiently as possible, using space with the greatest economy. In the bow, which was separated from the rest of the boat, I had stowed clothes, books, the telelens for my Leica, and my typewriter, all of which I knew I would not use during the trip. To give the canoe buoyancy in case she capsized, I also put in a few empty airtight containers. In the stern, in another compartment, I kept my food, spare parts and two small fluid compasses. In an easily accessible part of the canoe I stowed my drinking rations: four sixteen-liter demijohns of mineral water and two ten-liter demijohns of red wine. I mixed one part wine to three parts water, and this slightly sour mixture tasted good to me. Not only did it help drive off thirst, but the wine had the added advantage of containing easily absorbed calories. On the starboard side of my big compass I kept a brief case with sea books, and on the port, extra line and canvas; to starboard, above the brief case, was a small compartment, which held my logbook, a nautical almanac, a flashlight and photographic equipment. The small items I used continually, like sunglasses, suntan lotion, knives and can opener, I stuck into canvas pockets on the side of the cockpit. I did not shave and I had no mirror with me. I had no wish to look at myself during my torturous voyage.

At noon on the third day I sailed through the passage between the Farilhoes Islands and the Peniche Peninsula. Vegetation floated everywhere on the surface, and murres—small soot-black birds with white stomachs—crossed and recrossed my path. One might say they are the penguins of the northern hemisphere because of their white abdominal shirts, dark jackets and webbed feet, which are planted far back toward the tails.

I approached the mouth of the Tagus River, and for the first time since leaving Oporto, I saw big ships, some sailing toward Lisbon, others coming from the mouth of the river. I passed a ship from Hamburg and waved to the officer on the bridge. He waved back, little knowing that it was a fellow-townsman who floated near him in the strange craft.

With the last golden rays of the sun I sighted the roofs of Cascais and Estoril, home of exiled kings. I looked up at the hill terraces, where I had often sat in the warm winter sun, gazing down on the Alcántara pier, dreaming that one day I would make a voyage across the Atlantic, like the Portuguese sailors of old. And here I was, full of hope that this time I would succeed.

Night came. The long, thin lightfinger of the lighthouse at Cape Espichel circled above me. Occasionally, ships sailed fairly close to the *Liberia*, so close that I could make out shadowy figures on board. I was on a maritime highway now and could take no chances with sleep, but several times I found my eyes closing, and to keep awake, I resorted to singing and whistling, talking out loud to myself and to imagined friends. Nothing is more monotonous for the singlehanded sailor than keeping a night watch.

The next day, as I was taking my noon position, a cormorant, whose sparse feathers gave him the damp, naked appearance of a new-born baby, flew over the pirogue. This bird—whose name is a contraction of *corvus marinus*, sea raven—is a first-class fisherman, a skill that has made him of use to natives in some parts of the world. They force him into the water, a leather ring around his neck, and when he surfaces they remove his catch, which the ring has made it impossible for him to swallow.

On the fifth day I looked across white-capped seas to Cape St. Vincent, the most southwesterly point of Europe, where the dark eyes of large caves have looked for centuries upon shipwrecked sailors and sea battles. Columbus was rescued here, while he was still a young man, after his ship was sunk by a French-Portuguese fleet. From the steep, rocky cape I could hear the dull grumble of surf and swell, and then I came in sight of the rock of Cape Sagres, where, five hundred years ago, Henry the Navigator built his famous school of seafaring and exploration, which helped make his country a first class power for a brief moment in European history.

I sailed on with favorable winds. On the evening of the eighth day I stood before Casablanca, some 450 sea miles from Oporto. Calculating my actual sailing time, I found I had achieved an average speed of a little over four knots, a good performance for my canoe. I decided to wait until daylight before entering the busy harbor, although I knew it well from my work there three years before. During the night a stiff breeze came up from the northeast, an Ithacan counterwind—so called from Odysseus, who was often thwarted by counterwinds when he tried to sail into Ithaca—which blew me away from the harbor entrance. I sailed athwart for one hour, trying to get back, but the *Liberia* shipped so much water that I gave up, deciding instead to head for Safi, the second largest port in Morocco.

As I sailed past the heights of Mazagan on my way to Safi, I was surprised by a sudden stormy wind. It was useless to try to make further progress, so I put out the sea anchor and settled down for a nap. The *Liberia* behaved like a cork on the turbulent sea, rolling and bobbing ceaselessly, but I felt little of this inside and slept soundly, although I had to rouse myself every now and again to bail. The wind drove me back toward the east, but with a favorable breeze the next day I reached the heights again by evening.

The surf near Mazagan breaks at a depth of thirteen yards so that sailing there is dangerous and tricky. My entire attention was concentrated on keeping a safe distance between myself and the coast; I sailed with special care. A feeble breeze came from the west; it seemed wisest to enter the harbor of Mazagan immediately. Very, very slowly I approached the harbor lights. A terrible swell rolled under the canoe; I checked and rechecked my position to make certain that the distance from the coast remained at one sea mile. Through the darkness I could discern the white line of the swell breaking in the black night, a frontal attack of sea upon land.

Then, suddenly, I was caught. Giant breakers rose high above the boat and thundered at me on all sides. It was a sailor's nightmare; I had lost the passage to the harbor; I was trapped in the breaking swell. Small waves broke on the back of the mountainous swell, which raced at me with the speed of an express train. The sail was still set; each time the swell rushed by, the boom struck with such fury that my head was in danger. All at once I was faced with a moving wall of water; dwarfed, I crouched low in the canoe, held my breath, and then, with a roar, the wave struck. I pulled out the paddle. Would it help? I paddled like a man possessed; again I felt the boat shudder as tons of water poured over me. Was this the end?

Another breaker took me, like a toy, in its white claws, rolled under the canoe, and I paddled and paddled to get free. I was lifted into the air, but this time only the crest of the wave broke with a hissing sound as the *Liberia* settled back into the water. I struck desperately with my paddle to avoid the next breaker. A mountain of water hovered over me—would it break? Once again I was lucky, only the white top broke, partially filling the cockpit with water. I paddled athwart the swell, knowing I would not be safe until I had reached the deeper water that lay ahead.

The beam of a nearby lighthouse circled the skies, remote and unconcerned by my plight. In the distance the lights of Mazagan shone elusively through the darkness. I was alone, with no time to think out a course of action, with the paddle the only possible instrument of escape.

After what seemed like hours, I reached a spot where the swell was no longer dangerously high. I stopped paddling and dropped back exhausted onto the seat. My carelessness and stupidity had led me into that frightening crisis. In the dark I must have miscalculated my distance from the shore and so been thrust into the giant surf. I was wet through and the dugout was awash—flashlight and books swam in water, and the compass was nearly flooded. But my boat had come through undamaged.

In front of me a fishing boat loomed up in the night. I shouted to it, hoping to be towed into harbor, but only a dog, barking furiously, answered. So I set my course alone on the entrance to Mazagan. A rising tide and steady use of the paddle brought me at about midnight to the dock of the yacht club, where Arabs still fished. Clear moonlight shone on the old ramparts and minarets of the town and turned them to ghosts that hovered over the *Liberia* and the thick clay harbor walls. I spent an hour bailing the boat, changing into dry clothes and setting to right my possessions. Then I sat down on the washboard. I had no desire to go ashore until I had recovered from my terrible experience; I had to be alone, quiet, for my body still shook and I was drained of all strength.

In the morning I reported to the police, who graciously granted me a *permission de séjour* and simultaneously set a spy at my heels. When I came off the boat, my feet were swollen, and the next day blisters developed all over them. During the past eleven days at sea I had drunk a daily ration of seven fluid ounces of salt water and almost a quart and a half of other liquids. By the second day edemata had developed, which soon extended up to my knees. Otherwise, I was all right, except for extremely painful buttocks, where pustulae and then boils had developed. I had started out using a rubber air cushion, but found that it contributed to my discomfort for it pressed my wet clothes to my sore skin and cut off all circulation of air. The hard, wooden surface of the canoe became my favorite seat.

In Mazagan I bought thick canvas and lengthened the spray cover over the cockpit. I added fresh oranges to my provisions, and in two days I was ready to leave. Members of the local yacht club towed me out of the harbor, and after several hours of cruising, I sailed free of the treacherous breakers, which I remembered with painful vividness.

I set my course on the most northerly of the Canary Islands, Gran Canaria; with the help of trade winds, the *Liberia* made good time. I felt relaxed and happy in my progress and found time to admire the many Portuguese man-of-war jellyfish, that swam in the water beside me. I watched their pink combs that perch on top of light blue helmets and I thought how deceptively peaceful they appear on the surface. I knew from bitter experience, while underwater swimming, that beneath their surface beauty they carry poisonous tentacles, which can be extremely painful to the unwary or the unknowing.

The first night out of Mazagan there was a fresh wind. I tied the tiller, put out the sea anchor, and slept like a healthy baby. When dawn came, however, disaster jolted me out of my dreams; as I got set to sail again I found the rudder no longer reacted to the tiller. I suspected that the connecting split pins to the rudder pole had broken. I put on underwater goggles, which I had added to my supplies, jumped into the water, only to have my fears confirmed. There was nothing I could do at sea to repair the damage. I had to decide whether to try sailing on to the Canaries or to turn back to the North African coast. The coast was no longer visible, but as I knew it could not be more than twenty sea miles away I preferred to return to the nearest port, Safi. I used my paddle as a rudder, sailed only under gaff sail; luck and the wind were with me, and I sighted Cape Cantu, some ten miles from Safi, at noon. I hoped to reach Safi that day, so I sailed on, gripping the paddle with both hands, thankful that they were sufficiently calloused to prevent blisters. Evening came on too quickly to suit me; although the wind weakened, I continued sailing. At sunset I reached the bay, sailing directly under the huge rocks that guard the harbor entrance of Safi. I made three attempts to sail the *Liberia* into it, but an offshore wind thwarted me every time. I was finally forced to wait until the next morning, when I succeeded on the second try. I landed rudderless but without help, and my sense of achievement almost overshadowed my anger at a broken, useless rudder.

I moored the *Liberia* at the yacht club and set to work repairing my gaff sail, which had torn a little. Suddenly, from above me, a lady called down, "Can I help you in any way?" I explained my predicament to her and was delighted when she told me that her husband was a diver in the harbor and would be able to fix the rudder. The afternoon of the next day my new-found friend put on his diving suit and went into the water to unscrew the broken rudder. He made new split pins, welded the hinges to the rudder rod and had the blade made smaller. The next morning he towed me out of the harbor.

During the next few days the boat made good speed, but there followed several days of sultry calm, alternating with the merest breath of wind. My patience was tried to the utmost. I could do nothing but wait and follow my established daily routine, which was as regular and as punctual as the daily round of a banker. I took a bath every morning, I napped at noontime, I ate my meager meals according to the clock. Dolphins passed by; underneath the *Liberia,* small fish made their homes, and a long beard of green algae grew on her bottom and floated out in the water. Flocks of squeaking terns, sporting dark berets on their bright heads, and solitary petrels flew alongside. Occasionally I heard the snorting of whales; then the dugout would creep slowly forward for three hundred yards or so through their slimy, light brown excrement. Curious shapes were formed by the dung as it floated under the surface; once I made out the skull of a cow, another time I traced the outlines of a bare male femur. Often the wind blew it into long brown streaks, in which sea birds found their nourishment, reminding me of sparrows picking at horse dung on city streets.

I watched flying fish take agonizing leaps out of the water to escape their enemies. The water was full of plankton—millions and millions of microscopic plants and animals mixed with the eggs of larger sea life—and I spotted many swollen fish eggs in it. Small, dark jellyfish lay on the surface like tiny dust particles; the white shells of cuttlefish caught the sunlight and shone from the flat, lazy sea. On the fifth day out of Safi, the hours passed so uneventfully that all I wrote in my logbook were the words sunrise and sunset.

I do, however, remember one incident that occurred that day which I did not note in the log. I had nailed a horseshoe to the starboard side of the boat, a symbol of good luck that was constantly under my eyes as I sailed, but that day, to my annoyance, I tore my jacket on it. I recalled a sailor's superstition that wishes are granted when a loved object is thrown overboard; storms, they believed, could be calmed in this way; so I thought that perhaps, conversely, a heavy wind could be conjured up. I chiseled the horseshoe off the hull and, with many heartfelt wishes, threw it as far as possible into the water.

The next morning my wish was more than granted. The wind roared, the sea raged, and the pirogue staggered and reeled in mountains of water; I had just time to throw out the sea anchor and furl the sails. Big combers rushed over the boat. The hollow interior of the canoe magnified the sounds of the waves as they crashed against her; I felt as though I were sitting in a drum. Foam and spray seeped into the cockpit over my spray cover, and heavy breakers let loose floods of water that found its way to me. I bailed continuously, until my hands resembled a washerwoman's. There was no doubt that my horseshoe had outdone itself; this was a real storm. In spite of it I managed to sleep in short cat naps. By next morning the foam had abated, leaving only trade winds—still stormy enough to send watery messages into the cockpit. On the horizon the sea raged, giant waves rose up and broke up into boiling water. Petrels danced over the waves with such exuberance that it was obvious they prefer a heavy sea to a flat calm.

All day, because of the still-stormy wind, I sailed with the tiller tied and with two sea anchors out to keep from drifting south. Then when evening brought gentler winds, I reached for the tiller and with a shock found that it met no resistance from the rudder. I untied it and stretched out across the side of the boat to find out why. Foam ran down my neck through my open collar and dribbled down my chest, but I felt nothing; I was too intent on my problem. As the stern rose on top of a wave, I could see the upper hinges of the rudder trunk. They were empty. The rudder blade had fallen out of its hinges; it had developed an independent spirit and gone off to join the fish. I was furious, I screamed at the wind, at the empty air, anxiously waiting for the next wave so that I could examine the hinges again. I hoped against hope that my eyes had deceived me the first time, or that the rudder might have returned of its own free will to the copper hinges, like a repentant child to its mother. Whom should I blame for this disaster? The shipyard in Hamburg? The storm? My friend in Safi? Myself? It was my boat; I alone was responsible.

The ocean roared as before; combers rushed over the stern and into the cockpit, glancing off the deck like shot. Rudderless and helpless, I floated on the ocean, trying to keep up my spirits by losing myself in happy memories.

That night I put out two sea anchors and tried to sleep, but the noise of the storm made me restless. I awoke from a nightmare, bathed in sweat. Flashlight in hand, I crawled out of the cockpit and went to adjust the sea anchor. The stern rose and fell, now high in the air, then again completely submerged. With a grappling iron I reached for the line of the farther sea anchor; I found nothing. Damnation! The anchor was gone; I reached for the second one. Again the grapple found nothing. I pounded the seat in rage and frustration; I was going to be destroyed—rudderless and without a sea anchor—in the Atlantic in a bad storm. But I had not time to think; I had to act fast to find something that could go aft to replace the anchors. My knee happened to jab against the balloon sail. It would have to serve. I knotted a line twice around it and hung it into the water.

Morning came at last, and I found some sailcloth and sewed an emergency anchor. My hands still shook with fright from the night before. As the day lightened, a steamer came toward me; I was overcome by a strong, sudden urge to hail it. I had had enough—I was finished. But the ship passed about one sea mile to starboard, without a sign that anyone aboard was aware of my presence. I could see a face, very clearly, peering out of a porthole. The ocean between us raged and foamed. I waved with a white life belt. No answer. I climbed on my seat and waved again. No answer. The man had to see me; I would force him to notice me. But again no response. The boat rocked with such violence that I slipped off the seat into the cockpit. My whole body shook with desperation; at the top of my voice I screamed across the water at the ship, "Stop, stop, I can pay you for this. Please stop," and I held the life belt high in the air. But the tanker plowed on through the raging seas, while the face in the porthole gazed dreamily and sightlessly at the turbulent waters. I sank down in my seat, realizing that the sea was too heavy and too foamy for anyone to see me. For a moment my will to succeed left me. I was seized by a great depression. Should I jump overboard? Suddenly with a thunderous clap a large wave broke over the cockpit, knocking me over the compass. Water roared and rushed into every nook and cranny of the *Liberia;* my camera, watches and books were soaked. I bailed, I shook, I swore. I bailed again, prayed and bailed once more. When the work was finished, I found it had calmed me, although I was still too tense for sleep. I took five Dramamines, swallowed them dry —I had found that they act as a sedative. Then it occurred to me that the wine would help bring relaxation and sleep. I took hold of the demijohn and, leaning against the side, I drank deeply, drank and lay down on the bare wooden boxes—happy that I still had a berth at all. All I wanted was sleep, only sleep.

The next day the wind howled, and breakers hurled themselves like rocks against the hull of the boat. I bailed again and then crept into my cockpit. I was glad the tanker had not stopped to pick me up; my spirits had recovered. I finished sewing two new sea anchors and tied them both with plastic lines.

The furious face of the ocean did not change. Stormy

winds whistled; huge waves roared and rumbled. I whistled, too, and sang, and then swore at the unnecessary delay. I hunted for a jar of hard candies I had bought in Oporto because they were called "Good Adventure," but to my disappointment their taste was as bitter as my present situation. I hurt physically. From constant immersion in water, small ulcers had developed on my swollen feet and showed no inclination to heal; my buttocks burned as though a volley of shot had been fired at them.

The storm raged for four days and then the trade winds calmed, leaving a powerful swell. The *Liberia* had stood up well under the ordeal, but I had weakened. I had lost hope and been on the point of surrender. I took my noon position, then, three hours later, my longitude, and realized that I had been blown between the African coast and the Canary Islands, and to sail back against the currents was not worth the attempt. Dakar, the nearest good harbor, lay seven hundred sea miles south, which was too far; after some thought, I chose to head for Villa Cisneros, the capital of Río de Oro in the Spanish Sahara.

As I sailed back I came upon a locust, lying on the water but still alive. Whole swarms have been known to reach the Canary Islands from Africa. Another time a hideous cormorant paid me a visit, he must have been unusually courageous to have flown so far out to sea. To vary my diet, I speared and ate a dolphin. The wind lessened, and the Portuguese men-of-war sailed as fast as the *Liberia*. In a calm sea these jellyfish capsize frequently. They have to keep their combs wet, and in a flat sea they can do this only by listing heavily to one side; in a rough sea the comb stays wet from the spray, and the jellyfish rides out a storm with ease. Throughout the day I admired the constantly lively petrels, who were as at home in the sea as fish.

After three days of sailing, I heard the sound of breaking surf. Soon the white beam of the lighthouse at Cape Bojador pierced the haze. Later that afternoon, I looked upon a steep coastline that showed me only rock and sand. An unending stretch of beach where trade winds stir up the sands of the Sahara to an orange-red hell.

Rudderless, I had been forced to hold a paddle for fourteen hours a day; my left hand, once partially paralyzed during the war, was not suited to the task. It hurt from sunburn and its palm was covered with blisters.

After I neared the African coast, I was becalmed almost every day at noon; I found it difficult to decide which was harder to bear—a calm or a storm. When my nerves were strained to the utmost by the calm weather, I amused myself by teasing Portuguese men-of-war. I threw water over them, and they withdrew their pink combs, turning indignantly over on their sides. Plankton and fish filled the waters here; every day I met Spanish boats, for this is their main fishing bank. One time I had a sudden, and for me unusual, impulse to talk to someone, so I paddled up to a boat that lay at anchor, the crew angling over the side. I threw a line to them and a dark-skinned boy held it.

"Are you French?"

"No, *Alemán*," I answered.

"Do you have any cigarettes?"

"No"—my reply disappointed him. "But would you like some cans of food?"

Canned food is a luxury for these poor fishermen, and the captain, bearded and tanned, accepted quickly. "Sí, sí, señor." So I gave him some of my canned meat, wished him good luck and sailed on.

From the coast came the dull rumble of the surf. Dolphins, who had been my companions for so long, had left me now;

in their place were huge schools of herring-like fish. Flat calm weather plagued me; the sea was as expressionless as a death mask; only a light breeze filled my sails. My slow progress tempted me to put the dugout ashore and walk to Villa Cisneros, but when I looked at the sand dunes, I comforted myself with the knowledge that even poor sailing is preferable to hiking through hot sand.

Sailing without a rudder demands great patience; whenever the pirogue fell off wind I had to take in the main sail and paddle hard to return to my course. I woke up one morning—twenty days after my departure from Safi—in a thick fog, brought on by a cold coastal current meeting warm air from land. It was in part because of these fogs that the Romans named this part of the ocean the Dark Sea. At noon a fishing boat came up slowly behind me and with great difficulty sailed past. As it came alongside, we exchanged greetings and I found out that it, too, was bound for Villa Cisneros. When the captain heard we had the same destination, he threw me a rope, which I fastened to the bow of the *Liberia*. It was a lucky break for me, as, in any case, I would have had to be towed up the long archipelago to my destination. I climbed onto the fishing boat and arrived just in time for a cup of coffee. The next afternoon we dropped anchor in front of Villa Cisneros.

My dream of crossing the Atlantic in the summer months had vanished with my rudder, so I gave my provisions to the crew as a token of my gratitude. I had been afloat for fourteen days without a rudder; for two weeks I had steered a barely maneuverable craft with a paddle. During that time, my daily intake of sea water had been ten and a half fluid ounces, which I swallowed in doses of one and three-fourths fluid ounces six times a day, and now my feet and legs were swollen in spite of rest and exercises. I had proved to myself that there is no advantage to drinking salt water; it can, in fact, weaken a sailor's physical condition at a time when he needs all his strength.

Villa Cisneros is a place devoid of natural green; no trees, no shrubs, no grass break the monotonous yellow-brown of this Sahara outpost. A few deep wells provide brackish water, but drinking water has to be imported from the Canary Islands and is extremely expensive. As one approaches the chalk-white houses, one can see a little green, painstakingly cultivated by the owners, who are all, in one way or another, connected with the Spanish Army. The Portuguese once traded here for gold dust, hence the alluring name of the bay, Río de Oro—river of gold; but the river is as illusory as the gold, for it is only a long arm of the Atlantic that washes around the peninsula. The native Arab nomads, *moros*, live in blue tents strung out in straight military lines and swarming with flies and fleas. Dromedaries, goats and sheep loafed around the animal tents. The nomads are unusually hospitable, and often quite unknown faces and voices invited me to share their mint tea. I was amazed at how few children I saw, but learned that the infant mortality rate is exceedingly high.

One day during my stay I felt a longing to see the open ocean again and wandered down to the beach. Even there, I was followed by swarms of attacking flies; sand-colored beetles scurried away, their tails high in the air; dark, lazy beetles burrowed in the sparse vegetation. Trade winds suffocate the sea breezes here. Year in, year out, they howl around the corners of Villa Cisneros and stir the sand. It is an uncomfortable, unpleasant spot, and the only people who live there are those sent for professional reasons, and of course, the nomads.

After three weeks, I was rescued by an ancient mail

steamer, the *Gomera,* which carried me and the *Liberia* to Las Palmas in Gran Canaria. After we arrived, the pirogue remained for some time on the dock, suffering the expert inspection of curious fishermen and sailors. Finally, I was able to put her in a small shipyard for repairs.

I spent most of my time in Las Palmas on the beach or in the harbor, spear-fishing slimy cuttlefish, octopi and angry moray eels. I practiced my crawl so assiduously that my friends insisted I was planning to cross the ocean as a swimmer.

The known history of the Canary Islands goes back to the days of the Carthaginian explorer, Hanno, who mentioned them after a West African voyage. The name derives from the Latin *canis* and stems from Roman times when wild dogs roamed the islands. In the centuries that followed, Berbers, Phoenicians, Greeks, Italians, French and Spaniards contributed their part to the beauty of today's inhabitants. The island of Gran Canaria has been Spanish since Columbus' day, although Drake—and others—tried to wrest it from Spain. From the air Gran Canaria looks like an island volcano. It is cut through by deep gorges whose sheer sides fall away to the ocean. There is, nevertheless, such a diversity of landscape on this small island that one is forced to liken it to a miniature continent. Deserts with dromedaries, fertile valleys that produce tomatoes and bananas, high plains of wheat fields and pine tree groves give the naked mountains a friendly look. Date palms, coffee, papaya and pineapples are as native to the island as apples, pears, cherries and plums. Still lakes lie at the foot of steep mountain ridges and from an alpine landscape one looks down on tropical valleys. This is Gran Canaria, *continente en miniatura.*

My canoe lay in the boat yard where I was able to work on her in seclusion. Four weeks before my departure, I began my final preparations. I cut a square sail of six square yards, sewed a spray cover and attached an iron stave to the sides, over which the aft part of the cover could be drawn. By raising the level of the tarpaulin I hoped to keep away the heavy splashes of water and avoid the wind whistling through the cockpit. A new and stronger rudder was, of course, my major concern. A blacksmith made me a monster so large that a liner could have been guided by it. I had learned a lesson from the loss of my first rudder, and I took with me all sorts of spare parts, including a new mast of eucalyptus wood and a spare oar. One of the best sailors in Las Palmas happened to notice my gaff sail and remarked that it looked as fragile as newspaper. I was forced to agree with him, but I could not bring myself to part with all my old things. (The sail repaid my trust and stayed with me to the end. My spare sail played a part too as it served as a comfortable dry pillow throughout the voyage.)

The bottom of the boat I painted a reddish brown, the deck an ordinary white. Two days before my intended departure I moored the canoe in the harbor. I left that move to the last, so that barnacles and algae would have less time to grow on the bottom. But then my plans were upset by a storm from the south that swept the island. Palm trees bent like bows, banana trees were uprooted, masses of vegetation were carried to the sea by dirty flood waters.

After the furious south wind had worn itself out, it left the field to timid trade winds. The storm put me another three days behind in my schedule. My nerves were on edge. The day before my departure I knocked my alarm clock to the floor with such force that I had to throw it into the harbor as useless. I cut myself with a knife, and an attack of indigestion reflected my general unease. My last night in Las Palmas I moored the *Liberia* to lobster crates and slept on board. I trimmed the boat—a task that soothed me. On that last night my feelings were numb; no excitement and no fear stirred me. I knew that nothing would ever be as bad as the last storm without a rudder.

That evening, hot and perspiring from my work, I sat on the washboard to rest; shadows of fishing boats made a dark chain in the dead waters of the harbor; from the dock I heard the snores of the night watchman. It was the 25th of October, 1955. The hurricane season in the West Indies would soon be over. If the trade winds blew with their full strength, I could reach Haiti in two months. But past experience made me wary of planning ahead; only the Atlantic winds could dictate the length and the route of my voyage.

4 THE BIG JUMP

In the morning I sailed from the dockyard over to the yacht club, where I said good-bye to my friends. I had told them that I planned a voyage down the West African coast. I had kept my true destination secret in order to avoid comments and questions. One hour later I left the harbor of Las Palmas, alone and unnoticed; only the cathedral spire of the town and the naked mountains gazed after the canoe for a long time.

Outside the harbor the wind freshened; because I was under square and gaff sail the wake water moved in great whirls. Banana stems and torn cactus plants, remnants of the storm, floated on the surface. A fishing boat, apparently surprised at the foaming bow wave of my little dugout, came from the shore and ran beside me to test my speed. I was doing five knots. The boat came up under motor, and after a friendly waving of hands, she returned to her berth.

From the southeast of the island the Sahara of Gran

Canaria loomed yellow and inhospitable. The trade winds ceased. Pityingly, the lighthouse of Maspalomas looked down on the *Liberia,* struggling to sail free of the coast. I was sewing a button on the chin strap of my rain hat to tighten it, when I knocked my pouch of sail needles overboard. With all my needles gone I had no way of repairing torn sails during the voyage. My old standby, adhesive tape—which as a doctor I am in the habit of using on everything from torn shirts to book covers—would not stick to a constantly wet sail. Happily I had been foresighted enough to bring a spare sail along.

By afternoon I had cruised to the south of Gran Canaria. In the twilight a pair of porpoises snorted calmly and swam across my bow in slow motion. Mount Teyde, the volcanic crater of Tenerife, peered over its wreath of clouds like a Velásquez head over its Spanish collar. The clear moonlight gave the bioluminescent waters little chance to glow.

By next morning the current had dragged me back some miles toward the coast. Feeble puffs of wind were suffocated by a burning sun. The pirogue rolled in the dead water; with every swell, the boom menaced my nose, and my nerves were frayed by the aimless flapping of the empty gaff sail.

I took my sun bath, did my daily exercises and waited for the faithless trade winds. The islands would not let me out of their sight. To keep occupied, I took out my new Primus cooker, but in the high swell it would not balance on the bilge board. Instead, I squeezed it into a fruit crate, put a can of meat and an onion into my only cooking pot and lit it. Instantly it flared up. I jumped on my seat, grabbed my measuring cup and poured water over the flames; it seemed to make them worse. As they shot higher, they ignited the crate. Flames touched the deck. "I'll drown you before you burn me up," I shouted and, jabbing the prongs of the fishing spear into the crate, I hurled it overboard. Water suffocated the licking flames. I no longer had a stove; but I had chosen my food so that I could eat it hot or cold, and now the decision had been made for me.

In the late afternoon sun of the second day, I sighted the three southernmost Canary Islands. To the west, a lazy breeze made a fluted pattern on the surface; to the east, ship after ship steamed north or south. On the morning of the third day, a light breeze felt its way toward the African continent; but as I skimmed south I was surprised by a rain squall, which forced me to take in the mainsail. It was over in a few minutes, and I sat in a calm again, swearing at a poor wind which was forcing me into a dangerous steamship route. To the north, the cold, blue mountains of the Canary Islands taunted me with their indifference to my slow progress. On the fourth day I lost them behind the horizon; now at last the west wind freshened and the first combers fell into the cockpit.

To starboard, steamers passed me, headed for South America; to port, they passed bound for West and South Africa. For a single-handed sailor there is nothing worse than knowing he is caught in a steamship route. I, for one, felt as helpless as a seal on Broadway—clumsy and immobile in the face of oncoming traffic. I could achieve some measure of safety by using the system I had followed earlier—sleeping during the day, when the *Liberia* was more clearly visible, and watching alertly at night—but this is easier said than done. Hanging out navigation lights can only reassure a beginner, for it is pure chance if they are seen from the bridge of a steamer. Even here, in a coastal area which was crowded with fishing boats, not all of the big ships kept a watch on the bow. Although I was aware of the danger, there were several times at night when my head fell forward on my chest, and I dozed.

During my first week I ate only fruit and one can of meat a day. I had taken precautions against spoilage by wrapping my oranges and apples individually in paper, but on the fourth day some of my oranges had already begun to rot. Apples kept better. As long as the fruit stayed edible, I wanted it as a substitute for water and a protection against constipation, a condition hard to avoid on a concentrated diet and little exercise.

The west wind blew on the fourth and fifth day. This wind would have drawn a rubber raft to the Sahara. I managed to hold my own and stay in the steamship route, but still I felt a steady pull toward the inhospitable desert coast, of which I still had such clear and unpleasant memories. Rain squalls passed on either side of me, but always missed the *Liberia;* ships sailed by and an occasional fishing boat appeared over the horizon. I was constantly on the alert, and soon I began to long for some sleep. I searched the sky hopefully for some sign of trade winds that would blow me out of my misery. Nothing! No change! Dolphins flashed in and out of the water, chasing flying fishes; petrels fluttered in the high mountainous swell from the north; large shearwaters thrust themselves straight into the air, then shot down in the valleys of water, without seeming to move their wings; and, beside the boat, the light green shadows of fishes darted through the blue water.

On the evening of the sixth day, high white cloud banks sailed across the sky from the north; but at the water level, the wind came, as before, from the west. I remembered an old medical school teacher who used to say to his students, "Do not forget, gentlemen, when making a diagnosis, that you will find the usual more usual than the unusual." Would the trade winds finally obey this maxim and put an end to the unusual west wind? It seemed not. During the endless night the stars glittered, dolphins streaked through moonlit waters, but no breath of wind stirred in the sails. A caravan of steamer lights twinkled on the horizon. I had counted on the trade winds carrying me across the Atlantic. Would my third attempt also end on the sandy shores of the Sahara?

A full moon smiled down on the ocean and gave me enough light to write in my logbook. The next morning, at last, a timid trade wind blew from the northeast. Its harmlessness invited me to take a bath. I inspected the boat bottom and found the first small barnacles but no algae. That whole day the trade winds were so feeble they could hardly force themselves over the horizon, and on the eighth day I still dawdled in the steamer route near the West African coast.

The next morning, November 3rd, the monotony of my days was broken at last. In the many, many years I had dreamed of crossing the Atlantic in a small boat, I had always been certain that I would see something unusual in the ocean. Now at last I did. The sea was as flat as before; an even ripple lay over the surface; no foam was visible: the best weather for observation, the worst for sailing. A huge swell, rising close to thirty feet, rolled from the north; in its deep valleys fluttered Madeira petrels. I was watching them when all at once on the surface of the water I noticed an extraordinarily large brown shape. It disappeared into the next valley and then reappeared, this time in two distinctly separate parts, moving in the foam. On the fore part the sun's rays reflected against two big black eyes. I quickly reached for my camera; but as I was preparing to take a picture, the creature disappeared. Distance in the sea is difficult to estimate, it might have been as much as twelve hundred feet away; its length, also hard to gauge, looked to me to be at least double that of the *Liberia* and possibly as much as seventy-five feet. Though I saw no tentacles, I felt sure that what I had just missed with my camera was a rarely seen giant squid. Perhaps there were two together, which would explain the stretch of water in the middle. Their immense size has been estimated from tentacle scars left on the skins of sperm whales, who are their bitterest foes. They are shy creatures, like their cousins the octopi, and live in the deep regions of the water, where they can avoid their enemy the sperm whale. It was a bitter disappointment to me that the squid was so camera-shy. I waited for it to reappear, but it lacked any instinct for publicity. Someday, someone will be able to photograph this extraordinary ten-armed mollusk and it will be a real achievement.

On the tenth day the trade winds still husbanded their strength; a mere thimbleful of wind tried hard to push me westward. Yet only with difficulty could the *Liberia* outdistance the feathers, orange peels, corks and bottles that floated on the water. Water striders moved faster than I did. Around noon I heard the loud unrestrained snorts of a whale in the distance. The cameras were ready, and there it came . . . and this time I got my picture. It was Cuvier's beaked whale, a rare specimen, that came to cheer me up after the disappointment of the day before over the giant squid. The solitary whale headed straight for the boat, its dark body patterned by leopard spots, which emphasized its bright beaked head. No one knows whether these light spots on their skin are there from birth, are the result of bites from their own species or are produced by parasites, although I believe that the spots are too large for the last explanation to be the correct one. Then with great swiftness, my visitor turned back as if it had forgotten something, took a deep breath and disappeared into the depths.

Throughout the second week I sat in a flat calm; the wind seldom had enough energy to produce as much as one tired breath. Somehow, I had to pass the time during this waiting period: I paddled, I watched the birds and the fish and I indulged in daydreams. The surface of the water now looked as though a thin layer of oil or dust had been sprayed on it, but it was of course only plankton. I had my first visit from a tropic bird, who flew about my mast, trying to perch; he was unable to believe that I hadn't reserved a place for him. He tried over and over again and each time was frustrated by the small wind-indicating flag atop the mast. The only favor I could do him was to take his picture with my Leica.

On the eleventh day, with a slight wind behind me and the aid of the Canary current, I made my entrance into the tropics. Petrels and shearwaters flew listlessly over the interminable high swell from the north; they appear to need wind as much as a sailboat does, for without it they tire easily. They seemed as melancholy as I in a calm.

Small fish paid short visits to my dugout. On the fourteenth day, the first shark swam toward me; he loitered around the *Liberia* beating a steady tattoo with his tail fin against the cork pads on the side of the boat. I found that all sharks were curious when the boat moved slowly. Dolphins hunted their small victims and at times shot right out of the water in their eagerness. I threw them tidbits, and they rushed at them greedily only to turn away disappointed when they found they were not meat. Dolphins are confirmed carnivores, active day and night in their hunt for meat. During the night I could hear the hard beating of bird wings in the darkness. Before the moon rose, dolphins trailed long gray-white trains of bioluminescence through the water. Sometimes the canoe or the rudder touched a jellyfish or some other plankton form which lit up briefly and darkened again at once. Twice at night I caught glimpses of a dragonfly and a butterfly in the beam of my flashlight. Unfortunately, they did not have the presence of mind to take refuge on my canoe.

Every night now I was delighted by a display of shooting stars. On the sixteenth day a shearwater deposited its droppings in front of my bow and that night a comet swung down from the skies. Were they omens of good luck?

Throughout the long, dark nights I watched, entranced by the bioluminescence in the waters around me. Not enough has been written about this phenomenon of the sea, about the light the inhabitants of the sea make for themselves; for they are afraid of the rays of our sun and make their own sunshine. At night they turn themselves into spirits of light that haunt the surface of the water. They shine through the dark in a profusion of shapes and forms: spotlights, strip lights and floodlights are an old story to marine creatures. Taillights, halos, headlights are their natural inheritance. Nor is that all: plankton forms let other forms glow and shimmer for them; they catch bacteria which will light up at their command, or they grasp at lighted bacteria and bask in its rays. Jellyfish ornament their billowing skirts with shining bacteria and swim like ballerinas through the wet darkness. The fewer the stars that shine at night, the more friendly the light of the sea. Sometimes I found the nocturnal world of the ocean hard to fathom, but its very mystery lent enchantment to my lonely nights.

One night, as I watched the shimmering plankton, I thought I should like to taste it, at least once. It is, after all, the basic food of the sea. I hung out a net of the finest mesh, which had the effect of a sea anchor and made the boat groan over her additional burden. Because of this I left the net out for only an hour and then drew it up to examine my catch. Lighting my flashlight to investigate, I saw some sort of repellent vermin moving at the bottom. After a moment of hesitation, I took a spoonful and nibbled carefully. Immediately my mouth was full of an intense burning sensation. Scooping up a cup of sea water, I rinsed my mouth, and then smeared my lips with heavy cream; but the burning continued for hours. Since then, I have not fished for plankton—although I think it was not plankton that burned, but floating poisonous tentacles from a Portuguese man-of-war caught in the net.

On this trip, as on my two earlier attempts, Portuguese men-of-war often crossed my path. When the sea was calm—as it was during those first two weeks—I could see their light blue, pink-crested helmets across the mirrorlike surface of the ocean. Centuries ago, sailors from northern Europe knew that these jellyfish, swimming as they do in warmer waters, were an indication that their enemies, real Portuguese men-of-war, were close by, and they watched the horizon carefully. The helmet that these jellyfish wear is their sail. It is filled with gas, and when the sun shines strong and their silver-blue sails dry out, a nervous reflex forces them to turn on their sides to wet them. For the jellyfish it is self-defense, for when their helmets dry out, they wrinkle and collapse like empty balloons. It is a singular phenomenon of the sea, this Portuguese man-of-war—*Physalia*. It is made up of a community of tiny animals, of thousands of little individual creatures who have all developed special functions. Zoölogists call this type of colony of living forms *siphonophora*—a grouping of translucent, beautifully colored, floating ocean forms. On the surface of the water they show only their beauty, but under the surface the jellyfish carries a field of burning tentacles that turns the Portuguese man-of-war into a floating island of piracy and murder. Not all these tentacles perform the same function; some are there to kill and eat, while others are there to lay eggs and nurture them. This trail of tentacles, or streamers as they are sometimes called, can grow to the extraordinary length of ninety feet, and in a storm it acts as a steadying sea anchor.

On the high seas I watched a dolphin investigate a tentacle field; it hunted a fish that had taken refuge there and appeared to be immune to the stinging streamers. Not so the dolphin, who was burned, and quickly retreated from the field. I would not give one cent for the life of an underwater swimmer caught in a bed of these burning tentacles. I have often been stung by little pieces of these streamers that the surf has torn from the parent body and that have lodged on a coral reef; it feels like boiling tar poured on the skin, and even picric acid applied at once does not alleviate the pain.

After two weeks the ocean was still so calm that I could see, deep in the water, a swarm of dolphins mingling with their pilot fish. The shrouds had rusted and streaked the sail a pale yellow, so I decided to wrap them in sailing yarn. I was standing on the deck, working on this project, when I heard a loud splash before the bow and saw dolphins swimming for cover to the *Liberia*. My eye was caught by the passing shadow of a large fish. Another time, a periscope broke the flat surface, perhaps the arm of an octopus or the head of a turtle. It happened fast, and when I came closer, the periscope was discreetly submerged. The sea was alive; gelatinous masses floated everywhere, mixed with plankton and the excrement of whales. Only the trade winds had died.

I had to wait for my sixteenth day for a change. On that day heavy cloud banks approached from the north, followed by the longed-for wind. Shearwaters took a new lease on life, playing in the water like children and running lightly over the waves on their yellow feet until they were drawn into the air by the wind. Their cry resembled a goat's bleat rather than the song of a bird, and they did not charm me so much as the smaller stormy petrels; but they had the advantage of keeping me busy for a long time trying to classify them. Finally, after checking carefully in *Birds of the Ocean* by Alexander, I decided that my companions were the Mediterranean shearwaters, *Puffinus kuhli*, the largest Atlantic shearwater. Although they stayed near the *Liberia*, they showed no curiosity about her, seldom bothering to turn around and look at her. I did not find it easy to classify shearwaters and petrels from my small boat; if I had had weapons, I could have shot them and made accurate identifications then and there. Instead, I took color pictures and made my identification later.

Now the trade winds blew as they are reputed to, and I sailed westward. Squalls rushed over the sea and forced me to concentrate on handling the boat. I no longer saw much sea life; a few swarms of flying fish leaped into the air, fleeing dolphins, the sea birds became livelier, but plankton and water striders disappeared into the rough seas.

On the eighteenth day, the wind veered to the northwest —often a danger signal—clouds and sea merged and the horizon drew threateningly close. Crests of white foam capped the surface as the old trade winds and the winds of the new storm met and battled for control of the waves. At first the *Liberia* stuck obstinately to the old wave course, but with the help of my paddle I forced her into line. Moments such as these, when two wind directions meet head on with full power, can be extremely dangerous for small boats. I was relieved when the new winds took control of the waves and they became regular. Later that afternoon, the tornado —as squalls are sometimes called on the West African coast— had passed and the trade winds blew again. I was happy to see them despite the acute discomfort they caused. Whenever they blew with any degree of strength, I was wet—water dripped off my beard and ran down my neck, my sunglasses were blurred, and my drenched clothes clung to my body. The continual wet increased the soreness of my buttocks, which burned so intensely that I could hardly bear to put any weight on them.

On that same day—my eighteenth—I threw overboard the last of my oranges and apples; they had all rotted. Up to now the fruit had supplied me with all the liquid I needed. I was never thirsty and my intestines functioned beautifully. After the fruit ran out, I switched to a daily liquid intake of fourteen ounces of evaporated milk and a mixture of one and a half pints of mineral water and a bit less than a half pint of red wine, which I kept cool in a canteen wrapped in wet cloth. The raw onion I ate every day constituted my vitamin intake, and it was evidently enough, for I developed no symptoms of scurvy, such as bleeding gums, throughout the entire voyage. Columbus on his voyages took with him onions from the Canary Islands. Although he knew nothing of scurvy, he knew how to prevent the bleeding gums and loss of teeth that occur on long sea trips. Arab friends once told me proudly that long before the English gave limes to their sailors to prevent scurvy, Arabs always carried them on their dhows.

In addition to my daily ration of a fist-sized onion and a can of meat, I ate two small mouthfuls of honey at nine in the morning and at four in the afternoon. Honey has been prized for its restorative powers since ancient times and, like red wine, belongs to the old-fashioned family doctor's prescription for convalescents. I had brought it with me to give me added strength. Once in a while I ate the barnacles that grew on the *Liberia* or a triggerfish or dolphin I managed to catch. For centuries sailors have known that the sea can supply them with meat at any time. Recently the founder of the Oceanographic Museum in Monaco has added scientific weight to this belief by saying that shipwrecked people need not die of hunger if they have with them fishing tackle and a harpoon. Alain Gerveault came to the same conclusion after he had sailed alone from Gibraltar to New York in 1923, and the experience of those shipwrecked in wartime tends to confirm this theory. There is for example, the case of Poon Lim, a Chinese sailor, whose ship, the S.S. *Ben Lomond*, was torpedoed in the narrows of the Atlantic between Africa and Brazil. He spent 134 days afloat on a raft. His food supplies lasted only fifty-one days, and thereafter he subsisted entirely on fish, birds and rain water. He was eventually picked up by Brazilian fishermen and was still able to walk.

During my days of calms I had sailed through a sea that resembled bouillabaisse from Marseilles or a soup of fish eggs. Now that the trade winds blew, the soup had greasy waves. There were times when I grew alarmed at the violence with which the dugout rolled and lurched in the turbulent ocean. On the whole, however, I felt secure; my new rudder was strong, and the heavy following seas could do no real damage.

On the twenty-third day I spotted a white sail on the horizon. I assumed it was my friend Jean Lacombe, who had planned to sail from Las Palmas a few days after me. We had decided on a southern route and then a westward course along the nineteenth degree latitude in the expectation of a mid-Atlantic rendezvous. My heart lightened at the sight of the yacht and at the prospect of a brief respite from the discomforts of the *Liberia*. At nine o'clock, I prepared for my visit to Jean's boat; by ten, I had begun to wonder; by eleven, I was certain I had made a mistake. Jean's *Hippocampe* was smaller than the boat I saw and did not sail under a double spinnaker. As the yacht drew closer, I saw she was the *Bernina* from Basle. She came alongside under motor, and the crew asked me if I needed anything. After my "no-thank-you" they sailed on. Later I found out that the *Bernina* had made it in twenty-eight days across the ocean from Las Palmas to Barbados—a fine contribution to sailing, because she, too, had been hindered by little wind.

For two more days the wind blew with strength, and then on the twenty-sixth day it veered, first to the east, shortly thereafter to the southeast, and finally to the south. Soon the wind died, and the waves, subjects of the kingdom of the winds, flattened—only the ever-present high Atlantic swell remained.

The heart of the swell lies in the storm region of the North Atlantic. From there it is pumped southward throughout the ocean. The swell is dangerous only when it combines with rough winds at the beginning of a storm and throws up giant combers; now all was calm around me and I slipped into the water. My quick trip into the ocean brought to mind the story of the single-handed sailor whose boat had started on an independent course while he was swimming. Fortunately, such independence was not a part of the *Liberia*'s character. She was as attached to me as barnacles were to her. I did, however, take the added precaution of swimming with flippers to give me speed. There were many times during the trip when I raged at my *Liberia*, when she caused me more sorrow and anger than amusement or pleasure, but still I admired her. She was beautiful and she had character.

There was no wind from the east, no wind from the south and only an isolated, infrequent gust from the west. Occa sional local squalls swept over my shaky boat, so that I had to handle her carefully, but not until the twenty-ninth day was there any real change in the weather. On that day, a menacing wall of cloud gathered its forces in the west, swept quickly toward me with ominous aspect. I took in my sails, put out the sea anchor and took refuge in my hole.

    I was awakened by a noise like thunder. My first sleepy reaction was that I was being called upon as a doctor to help in an emergency; then I realized I was in the *Liberia* in mid-Atlantic. A shark had wakened me by banging against my boat. Like a comfortable citizen, disturbed at his noonday nap, I was enraged at the shark's lack of consideration. I pulled out my Leica and took his picture. He was not impressed. He slunk along the starboard side, a little on one side and inspected my boat. He disgusted me with his wide, cynical mouth and small pig eyes, which seemed the essence of greed. He was nearly fifteen feet long and such a giant that he appeared heavier than the pirogue. I pulled out my movie camera, pressed the button . . . but nothing happened. I tried again with no result; the camera had rusted and filming for this voyage was finished. The shark slunk off, taking with him my following of little pilot fish. As a farewell gesture, he splashed the bow with water with a flip of his huge tail. I watched my faithless pilot fish dart nervously around their new host, thinking to myself how human they were with their characteristic of attaching themselves to those that can give them the most. When I began to steer again, I found the rudder did not work. The shark's blows had broken both port-side cables that controlled the steering. If my rudder had been less secure, the blow might have had more serious results. I replaced the torn cables with strong wire, an easy job now that the storm had passed, and no wind blew. Later I strengthened the other cables, in case another shark should decide to use my rudder as a target. From now on I gave up steering with my feet, as I could only do it in calm weather when I had no need to. In windy or stormy weather, I had to sit on the washboard and hold the tiller to balance the boat. To keep my boat as dry as possible, I closed up the small holes in the hull that I had made for the foot-steering cables.

Still the wind came from the west. I decided that the pirogue would make better time on its own, riding the westward current, so I let the sea anchor take over. The halyard to the square sail was frayed, and to keep myself occupied, I decided to replace it. I tied the ends of the old and new halyards together, but the old one broke near the top and fell to the deck. What was I to do now? I couldn't take down the mast because its base had swollen fast in the deck. If I climbed the mast, the pirogue would capsize. Finally, I came up with a solution. I tied the halyard loosely to the grappling iron and wrapped wire around its end to stiffen it. Then, standing at the base of the twelve-foot mast, I reached up and threaded the line through the block with ease. But to my irritation, I found I could not free the iron. "Patience," I counseled myself, "there is more than enough time." The canoe bobbed about in the high swell, but after several hours of painstaking, steady concentration, I finished threading the halyard through the mast block; I pushed with my left hand and drew it through with my right. The job, performed under a broiling sun, had taken every ounce of will power and patience I possessed, and when I had completed it, I rewarded myself with a double ration of wine and water.

I was now close to the middle of the Atlantic. Only the swell disturbed the absolute calm of the ocean. I sailed through a gigantic aquarium of floating plankton, of water striders that zigzagged around me, and of triggerfish that nibbled at my barnacles.

None of the fish I saw gave me greater pleasure than that funny creature the triggerfish. I had my first visit from triggerfish when I sailed into the tropic of Cancer, and they consoled me for the lack of wind. I had been watching, with mounting irritation, my sail flap idly to and fro, when my attention was caught by these little fish popping out of the water. They waddled around the *Liberia,* matching her clumsiness. Triggerfish do not hunt other fish, but subsist on parasites or any flotsam and jetsam they find. From my boat they had luxury meals of barnacles, but their usual source of food is plankton. They followed me all the way across the ocean, waddling along behind the canoe in calm weather, like baby ducklings behind their mother.

Alone and bored in mid-Atlantic, I talked to the triggerfish, at times in friendship, at times in anger. My attitude toward them depended on the weather: "Good morning, my friends, and how is everything with you today?"—but then, when the trade winds refused to blow, I screamed at them, called them thieves for stealing the barnacles from the *Liberia* and ordered them to leave. That they neither heard nor understood did not bother me; it was enough for me that I had someone to talk at, I did not need an answer. Sharks and dolphins grant triggerfish a safe-conduct pass and never attack them, but I was not so kind and killed them with my spear whenever I could. When they are in danger, triggerfish push up their dorsal fins—it is from this trick that they derive their name—but they do not fight back. If they are fished out of the water, they turn pale and give one shocked grunt; because of this sailors have nicknamed them "old women."

I did not kill for the sake of killing but to enrich my

knowledge of the sea; all of us who have sailed the Atlantic can add to man's knowledge, if only in small details that may eventually form the basis of a scientific formulation of our combined experiences.

Triggerfish meat has reputedly poisoned people, so I approached mine cautiously. I ate a little less than an ounce the first time, gradually increasing the amount so that within a few weeks I was able to eat a whole one-pound fish. I suffered no ill effects. Fishermen of the Caribbean islands know that poisonous triggerfish are caught on one side of an island and harmless ones on the opposite side of the same island; it seems likely, therefore, that these fish are safe to eat unless they themselves happen to have eaten poisonous plants or fish. I know that barracuda, which we ate as a delicacy in West Africa, is considered unsafe to eat in the Antilles or the Pacific. Also, the fact that poisonous fish are less likely to occur on the high seas gave me confidence in my experiment with triggerfish meat.

Although my primitive way-of-life on the *Liberia* had coarsened my taste buds, I could still detect a difference between the flesh of a dolphin and that of a triggerfish; the former is tender and tasteless and the latter is tough, salty and slightly bitter. Of course, since the loss of my Primus stove at the beginning of the voyage, I had to eat everything raw. I had no aversion to raw meat; as a matter of fact I found that the organs tasted better raw than cooked. Sometimes I added flavor to my fish with salt water in the Polynesian manner.

Every speared triggerfish had in its stomach barnacles from the bottom of the boat, most of them still bearing traces of the supposedly poisonous red paint. Sometimes twenty or more congregated under the *Liberia,* excitedly nibbling the barnacles, then surfacing with a quick, explosive "tsh-tsh," either a belch or an expression of pleasure, before they returned to their meal. After I had speared one, his companions craftily stayed close to the rudder or bow where I could not reach them.

During the first seven weeks on board, while the calm weather lasted, I spent a great deal of time in the water with goggles and flippers. I was careful to keep a look-out for sharks, but they were busy elsewhere; I substituted as host to two small shark-following pilot fish. Groups of these nervous little "yes-fish" were often to be found swimming around the boat, changing daily, so that one day my following would be tiny blue-striped fish and the next larger green pilot fish. There was only one faithful pilot, who stayed with me for fourteen days, making his home close by the rudder. When I swam in the water, he entertained me with his antics, but was careful to stay at a safe distance.

I had heard many stories in my life, about the ravages of man-eating sharks. I had only half believed them on dry land, but once in the water—and vulnerable—everything seemed possible. As I swam around the *Liberia* one day, a dark shadow on the African side of the boat caught my eye, and I drew myself quickly up from the fearful regions of the water into the shelter of the cockpit.

On the thirtieth day I had the pleasure and excitement of seeing my first school of small whales. I watched them surface and blow water-laden air from their lungs. Even in the humidity of the tropics this warm water vapor is visible to the naked eye. Each variety of whale has its characteristic spout, which enabled whalers to tell them apart. When whales surface from a great depth, they take several shallow, quick dives to enrich their blood with oxygen, and this is exactly what my group of little whales were doing. They took deep breaths and disappeared for several minutes.

Large whales can stay under the surface for over an hour and dive to the depth of half a mile or more, but, like men, they are vulnerable in the water. I remember an incident that occurred in 1932 off the coast of Colombia, when a sperm whale got caught in a telegraph cable at a depth of over three thousand feet and drowned.

The next day the wind from the northwest was still slight, although clouds, pulled into long strips by the wind, appeared here and there in the sky. A painful boil had developed on my neck and caused my glands to swell. Smaller boils appeared on my thighs, due, I think, to dampness, and I had to take aureomycin for three days before they disappeared.

During this calm weather, that tried my patience and hindered my progress, I kept to a set daily routine that helped pass the time. Every day, at midday, I took my noon position—next to staying afloat, the single most important job on a boat—and then put the sextant, that precious instrument, very carefully back in its wooden box. Sometimes splashes of water had wet it, and then it had to be dried in the sun before I packed it away again. It took me only a few seconds to figure out the latitude and then another few minutes to arrive at the longitude, by means of the time difference between Greenwich time and my time. The accuracy of my calculations reflected the condition of the sea: the higher the waves, the less accurate my results. In calm spells my logbook frequently noted the same latitude for several days. With my navigation completed, I took a sun bath and spread out around me my brief case, books, cameras, clothes and blanket; even my valuable onions shared the sun with me. Despite these precautionary drying-out periods, my cameras rusted.

The sky was seldom the azure blue it is over the Mediterranean; in fact, the intense humidity often turned it into a leaden gray. On windless days the rays of the sun pierced the overcast with tropical intensity, and I sat with a wet towel draped over head and shoulders. Still, I sweat more than was good for my health. My skin began to show the effects of dehydration, and on dry days I increased my ration of liquids. Although I was not necessarily more thirsty than usual, I knew that thirst is not an accurate measurement of a body's need for liquids, and under conditions like mine, one has to drink more than one thinks necessary.

At four o'clock I ate—a can of meat and an onion. To relieve the monotony of my diet, I served myself beef one day and pork the next; I always drank an apéritif of wine and water. In the evening, as the hours dragged by, I sat and licked at my weekly ration of honey. On days that brought discouragement and loneliness, I cheered myself by eating a whole pound of honey at one sitting. With the setting of the sun, the temperature would drop and I would put on more clothing; when the wind was strong I even wore a cap for protection against the cold. In the last light, I would write my final entries into the logbook.

My narrow canoe rolled and yawed so badly that I usually took in the gaff sail and went under square sail at night. The sun pointed his last golden finger into the sky, the evening star appeared and there—suddenly—the *Liberia* and I floated alone in the tropical night. The air cooled, and no matter how much clothing I piled on, I shivered and rheumatic pains shot through me. Tropical nights are long, and at sunset I faced a long period of inactivity; on my tiny boat there was nothing to do but sit and think. These night hours, therefore, seemed doubly long, lonely and frustrating; I could only wait in enforced inactivity for them to pass. But I delighted in watching the drama of the clouds; I traced pictures in their shapes. I watched as dogs and horses were chased across the sky by gnarled and twisted apparitions from Hades, to be replaced in a few minutes by magnificent cloud trees. Then, as a mass of clouds streaked across the sky, shapes, pictures and outlines would dissolve into pitch-black night.

During the night my subconscious—constantly aware of potential danger—exaggerated everything: shadows of great waves menaced me; breakers loomed high, frighteningly large; and the *Liberia* sped too fast through the water. When the combers overwhelmed me by their size, I did not sail at night; instead I threw over the sea anchor, took in the gaff and square sail and lashed them to the deck. I drew my spray cover over the iron arch I had built to keep the wind from blowing around my bare toes and crept onto the packing cases, put my head on a seabag and slept; that is to say, I napped in short intervals, depending on the weather outside. Even in sleep, I was alert to possible danger and always awoke at the first splash of water into the boat, as a mother wakes at the first cry of her child. When the wind was favorable and allowed me to sail I never slept more than five and sometimes only three hours. When the waves were fierce or the wind contrary, I managed twelve hours of sleep —with breaks for bailing, of course.

I knew from past experience that my life depended on sleep, that without it I was incapable of being sufficiently alert for safety. My attack of delirium in the Gulf of Guinea had taught me that lack of sleep is often the biggest single danger a lonely sailor faces. I was determined not to allow it to happen to me again.

Everything I did on the *Liberia* depended on the wind. When no progress was possible because of stiff winds or a calm, I slept until dawn. With favorable winds, I was up at two in the morning. My first morning chore was to bail. Then I hauled in the sea anchor, whose line was often covered with particles of bioluminescent plankton. The next step was to set the square sail. This was no easy job in a strong wind; the sail had to go around the shrouds in the forward part of the boat so that the wind would suddenly fill the six square yards of canvas. Then the canoe would respond by lying over on her side. Sometimes I was afraid I might capsize, and often I was forced to drop the sail into the water to right the boat. A near miss like this in the middle of the night always startled me, and I usually waited until daylight to hoist the gaff sail. I had resolved at the beginning of the voyage not to lose my temper or shout when something went wrong, but often after a misadventure I had to remind myself again of my resolve.

During the morning, my mind and body refreshed by sleep, the hours passed quickly; I observed the life of the sea around me, daydreamed and did my exercises. I did knee bends, massaged my legs and executed a sharp—albeit stationary—military march on deck. At nine every morning I drank a fourteen-and-a-half-ounce can of evaporated milk.

A whole month went in this fashion: small squalls, calms broken by an occasional good wind. I waited, hoping for a good wind that would take me across the Atlantic and, at last, on the thirty-second day, November 27th, the trade winds came to my rescue. The red-billed tropic birds and petrels took a new lease on life. Mediterranean shearwaters paid me their last visit; I saw them no more for the rest of the trip. In the evening a shark swam close to the pirogue, dousing me with a flip of his tail. In the pale moonlight he looked as evil as his reputation. Perhaps he sensed my revulsion, for he said his good night quickly.

The trade winds blew vigorously for two days. Night sailing became dangerous, so I put out the sea anchor and went to bed with the start of darkness. The waves pushed me forward. I thought of them as my persecutors; powerful and mighty, they attempted again and again to change the course of the *Liberia*. They were especially dangerous and importunate going down wind; then the canoe yawed and rolled, moved from starboard to port, luckily never making a real sideways jump. I knew the *Liberia* well by now and would not tolerate any tricks from her. I kept my hand steadily on the tiller, ready to yield or to oppose as the occasion demanded. I allowed her very little freedom because I knew she could not handle it.

I remember a completely unforeseen moment of danger that occurred during those days, as the sea roared and stormed and foam danced on the surface. Swell and wave met, and giant mountains of water rolled under the pirogue, broke head on with a loud roar and left a flat foam-scarred surface behind them. I exulted in my speed and in the frothing wake of the *Liberia*. The boat skimmed over the seas, and I had her well in hand, I thought, as she sailed flat before the wind. I was careful not to look back. The noise and froth looked more dangerous behind me than ahead of me. So I sat still, keeping my eye on the square sail and on the passing seas. Then it happened. A giant sea heaved the stern higher—higher—higher, and the bow dug deeper—deeper into the water. I clutched the tiller. I held my breath. The sea broke aft with a crash and the *Liberia*, for one frightening instant, stood on her head. Then she righted herself, and the sea passed under her. I was safe, but the fright that had gripped me left me weak and nauseated. My stomach refused to settle down. I knew how easily a small boat can capsize when her bow is too buoyant. To calm myself I threw out the sea anchor, took down the sail and breathed deeply.

Although the average height of the waves was only eighteen to twenty-four feet, occasional larger ones, measuring anywhere from thirty to forty feet, rolled by, and when these mountains of water broke, the resulting roar chilled and shook me.

I had been afraid that, as the voyage progressed and I consumed my food stores, the *Liberia* might become too light. As a precautionary measure, I had put all empty bottles, jars and cans in the bilge water behind the aft seat. The constant and terrible roll of the dugout made the empty containers glug and gurgle as the water in them rose and fell. To me they sounded like the voices of men and women; they shouted and whispered, laughed and giggled, tittered, coughed and mumbled. Their voices became so clear that I finally joined in the discussions. But no one seemed to care very much for my opinions; they laughed at me and kept on in their own strange language. Hurt, I no longer added my voice to theirs; I was convinced I had been wronged. Later I had my revenge. I took the water out of the stern bilge and thereafter only an infrequent sad, quiet sigh floated up to me.

The trade winds that blew at the end of my first month at sea brought with them quick, scurrying squalls. Menacing banks of clouds, reaching down into the horizon, would darken the sky and hurl buckets and buckets of water at me. I began to dislike the sight of these clouds that portended rain, not because I dislike rain, but because it fell upon us with such force that the pirogue lay over to one side. I was always forced to take down the sails. I hated and dreaded these mid-Atlantic squalls that exposed the weaknesses of my boat, leaving me impotent in the face of their brutal strength. Not a day went by that I was not menaced by one;

sometimes they enveloped me; at other times, in a kindlier mood, they left me out of their play.

On the thirty-fifth day red-billed tropic birds flew over me ceaselessly. The next day only one came, and then for seven days, they left me alone. I missed them. Their unabashed appraisals of my boat always gave me pleasure; they never flew over the *Liberia* until they had given her a careful scrutiny. Then they swooped over me in curves, yip-yapping back and forth to each other. I was sorry I could not understand their comments on myself and the boat. These birds, who limit their flying range to the tropics, were to me the most charming of all sea birds. I could recognize them from afar by their high flight and the steady pigeonlike flapping of their wings. As they approached, I was even surer of their identity; for they sport long tail feathers, and a salmon red touch on their wings, a lobster-colored beak, a black band near their eyes and black stripes on their wings—easily distinguishable markings. They are the same size as gulls, but unlike them, they fly high above the water's surface, flapping their wings in untiring energy. Only once did I see a "son of the sun"—as Linnaeus, the eighteenth-century Swedish botanist, named them—rest on the water.

On the thirty-seventh day, the wind veered from northeast to northwest. Such a change could mean danger, and I put out the sea anchor to give me stability. The sea roared, water rushed into the boat and I bailed frantically. I worked hard and worried hard all through the day. I was concerned not only about the weather but about my health, for my intestines were not functioning as I knew they should. I swallowed half a bottle of mineral oil in an attempt to solve this problem. In vain. Then I realized that my concentrated food did not contain enough roughage; my system knew exactly what it was doing, and I should stop interfering with its functioning.

That night the seas calmed and I was left rocking in a high swell. Triggerfish again harvested barnacles from the *Liberia*'s bottom, a new variety of shearwater performed in the sultry air, and water striders tried to scramble from the corkwood pads onto the side of the dugout. I came across these little insects of the sea quite often during the ensuing calm days. They are common, even in mid-Atlantic. Their ability to stay on the surface of the water reminded me of childhood experiments when we floated an oil-covered needle on top of water.

The *Liberia* and I had now been afloat for forty-one days and the time had come to take stock of my situation. Back in Las Palmas I had estimated—and thought myself generous—that forty days would see me in the Antilles. But now I knew my generosity had been inadequate, for here I was, still bobbing around in the middle of the ocean. For seventeen days I had had contrary winds that had done their best to push me off my westward course, and ten days of only slight wind had afforded an excellent opportunity for a closer acquaintance with the life of the sea, but had brought me little nearer to my goal. The Atlas of Pilot Charts had promised me trade winds, but as I found out later from other sailors who made the crossing at the same time, the wind conditions that particular year were not normal. They forced me to relinquish my plan of a non-stop voyage from Las Palmas to Haiti and to decide on a closer goal. I chose St. Thomas in the Virgin Islands. I did not feel easy in my mind about the rest of the voyage, for I was haunted by a fear of further contrary winds that would strain my calculations to the utmost. But there wasn't anything I could do to alter whatever the weather had in store; certainly brooding and worrying would not get me one sea mile closer to the Americas.

I vented my anger and frustration at the delay by spearing dolphins, and cutting them up. At one point I had tried catching them by throwing out a metal fish lure, but the dolphins were far too wise to bite. So I invented another game to play with them that would occupy my mind; I cut up a speared dolphin, strung its vertebrae on a nylon thread and threw it into the water. Within seconds, its living brethren had devoured the meat. I hung out more and the dolphins followed me, like a swarm of begging children behind a tourist; I held a piece in my fingers over the water, and—snap—the most courageous bit it right out of my hand. I kept on at this game; my canoe followers were insatiable. In their excitement their dorsal fins moved up and down, and they fell furiously on their prey, oblivious to all danger.

The trade winds blew unenthusiastically for the next few days. Little drops of fish blood on the tarpaulin smelled so badly in the tropical heat that I had to scrub it thoroughly. I enjoyed a cooling bath in the waters of the mid-Atlantic. I watched tropic birds fly by and wondered if they could be my first messengers from America. I wasn't sure. They could have followed me from the Cape Verde Islands or they could be welcoming me to the Caribbean. As their numbers increased in the following weeks, I began to be certain that they were messengers from the New World.

On my forty-sixth day I was shaken by an untoward occurrence. The trade winds blew stiff, and I had not dared hoist my sails during the night. At intervals I crept into the cockpit and bailed. Suddenly, only a few hundred yards away, a great lighted steamer came across my course. The sea anchor aft with some thirty yards of line plus the length of my boat made a vulnerable "beam" of about thirty-eight yards. It was a bad moment for myself and the *Liberia*. The danger passed, but it left me in a brooding mood. I thought with envy of the crew—of their dry clothes, their foam rubber mattresses, and of their dreams of their next shore leave. In contrast, my *Liberia* offered no physical comfort and demanded a constant watchfulness; at a moment's notice my whole being had to be ready to do its utmost to keep her from capsizing. The steamer, plowing powerfully through the water on its way from New York to Capetown, so solid and secure, depressed me. But I knew I was there—alone—of my own free will. Even during my worst hours, when I felt most deeply discouraged, this knowledge helped me.

Again I was persecuted by squalls. I would not have disliked them so much if I could have caught their rain, but my boat rolled too much to make that possible. One must have a beamier boat to be able to collect rain water. Once (unfortunately, only once) I found two tiny flying fish on my deck in the morning. My boat was also too narrow for this excellent source of free food. Occasionally other flying fish "flew" onto the deck or sail during the night, but they always fell back into the water.

My hands were in bad shape; the calluses had worn off and left them sore and open. I was forced to wrap them in a towel or a sock before I could touch anything. My skin mutinied, too, by breaking out in boils. Two of them made it impossible for me to sit in comfort, and one on my thigh made my foot swell. For three days I took two aureomycin capsules every six hours. My physical condition and my concern over the weather led to moments of deep depression and frustration. During my worst moments I found myself singing—songs from my childhood and student days; when the sea was rough, they turned into monotonous chants. I often heard myself repeating, for hours, the opening bar of one song. I had come to a point where the world beyond my horizon no longer existed. Here at sea other forces ruled my life; I was alone with God, alone with nature, alone in remoteness and solitude. Yet I experienced none of that sense of loneliness—that "cosmic feeling of solitude" as it has been called—that can afflict us in the presence of other human beings. Loneliness weighed on me no more than it does on a healthy child. Like a child, I peopled nature with my friends. I talked with tropic birds and addressed remarks to water striders skating over the surface. Conversing with nature calmed me.

Now the trade winds blew briskly and night combers raced by the *Liberia*. Even my stomach churned at times. Squalls darkened the foaming waters, while banked clouds scurried across the sky in all directions. My fish deserted me in the heavy seas.

In those days of stiff wind, the setting of the square sail required a great deal of patience. A couple of times I had to let it fall into the water in order not to have the canoe capsize. Despite the high seas, I took my noon position; I had to be sure that I was not drifting too far to the south, where trade winds blow up storms that last much longer than those in the north.

On the morning of the fifty-fifth day my flashlight went overboard. I had put it inside the life belt which was lashed onto the deck in front of me, but at a roll from the *Liberia*, it fell into the water. The speed with which this occurred made me realize, with a sinking sensation, how easily it could happen to me. With my flashlight gone, I had to hoist the sail in darkness without being able to check the lines. The results were unhappy; the jib fell into the water, its line broken. Instantly I threw out the sea anchor and cut down the sail. The new blocks I had put on the mast in Las Palmas, to save weight, had frayed the line. I sailed that day with only a gaff sail; but the best gaff cannot replace a jib, and I found the sailing very tricky. Despite the trade winds, wind and waves came at me from more than one direction, and the high swell, surging from the north, combined with the wind from the northeast and rocked the canoe athwart. Then a sudden breeze would blow from the southeast, throwing the boom from port to starboard. It was dangerous sailing that demanded constant and exhausting care and watchfulness from me.

The wind died down the next day. I was then able to replace the broken line. My previous experience with a similar task helped me, and I did it much faster this second time.

In mid-Atlantic I had had petrels with me constantly. Now I was leaving them behind. At twilight on the fifty-eighth day, in a symbolic gesture of farewell, the last one flew against my sail. I never saw any again, although they are not unknown in the Caribbean. Early on the fifty-ninth day, as I was setting the sails, the jib line ran out of my hands. It hung in the air, and as I climbed to the bow to catch it, I looked up and saw my first "Americans," two frigate birds, circling in the air above me. They were the first sure messengers from the continent that lay ahead. I knew that these birds are seldom seen more than one hundred miles offshore, and their presence raised my hopes. Was it possible that I might reach Antigua by Christmas?

In the soft light of early morning these frigate birds or man-o'-war birds looked even blacker to me than their plumage and their reputation warranted. Like eagles, they circled menacingly in the sky, ready, when they spotted their prey, to plummet to the surface of the water, with their fork-shaped tails spread, grabbing for their food with long hooked beaks or sharp claws. I watched them hunt dolphins and catch flying fish that had been chased into the air. The long-winged frigates are sea birds that cannot rest on water, for they need an elevation to rise into the air. They generally spend their night on shore, where they sit, vulture-like, gathering their strength for the next day's hunting.

At noon on the fifty-ninth day the trade winds failed. What little strength they had appeared to derive from occasional rain squalls. This delay as I was nearing my goal became almost insupportable to me.

On December 24th the winds remained unchanged. Here I was in the most accurately predictable weather system in the world, but, as elsewhere, the forecasts proved false. Morning brought eight frigate birds, and at noon I heard the drone of a plane. I knew then that I would reach Antigua the same day. My Christmas present to myself. I sailed on with renewed hope and vigor until late that night. But no welcome lights rose through the darkness before me. I began to doubt the accuracy of my chronometer. I wondered if my longitude were wrong. Still, I sailed on. At two o'clock the next morning a squall brought thunder and lightning; at three, the storm ended, and I saw two lights rise and fall behind the high swell. Steamers? Fishing boats? I could barely see them through the enveloping darkness. Were they, perhaps, the two lighthouses of Antigua? According to my noon position, they lay before my bow. Then, at four that morning, I saw lights that shone from a high point of land and I knew I had reached Antigua. Behind those lights, I thought to myself, someone is celebrating a tropical Christmas; so I set the sea anchor, took in the sail and looked around for some way of celebrating on my own. All I had left was canned milk and honey, but in my jubilant mood it was enough. I was the first to cross the Atlantic alone in a dugout canoe, and I had achieved my goal at Christmas. I could not imagine a more welcome present.

I drank my cocktail of milk and honey in my dugout paradise, my thoughts turning to Christmas, to Antigua and to the people who sat behind the bright lights while wet trade winds hurled through the night. It was now five in the morning. Were they still awake in their homes? At least their lights shone at me through their windows. And it was America and Christmas morning. So I began to sing "Silent Night" and "O Tannenbaum."

I listened for the welcome sound of surf, but the trade winds carried it to the west. I left the decision whether to sail north or south of Antigua in the hands of the current. It took me south, so in the early dawn I set sail for the southeast cape of the island. It lay before me, cloudless in the sea. Then slowly the hills collected a white cap of cloud and anchored it there. These island clouds are visible for many miles, for they rest, immobile and heavy, on the mountains; unlike the light, fast, white trade-wind clouds, they are not easily driven off by the winds. Only a bad storm has the force to move them.

As the sun rose on the chain of Caribbean islands that stretched before me, I noticed a curious phenomenon: these islands were linked by clouds that ran from Barbados to Antigua, and from there to southwestern Montserrat, which suffocated in heavy clouds.

I sailed close to shore, hoping to see the famous harbor where Nelson had anchored, in hiding, for several years. I was unable to catch a view of it, but I admired the green hills, pretty bungalows and white beaches that lead down to cobalt waters. The sight—my first sight of land—awoke in me a mood of irresponsibility; I felt childishly gay and playful. Around the *Liberia* floated pieces of wood, put there by fishermen to mark their lobster pots; to me they symbolized human life and activity, a sign that I was no longer alone in the world.

As I sailed southwest of Antigua, the wind brought me the smell of balmy land air, reminding me of dry grass and hay. I was suddenly overwhelmed by a longing to set foot on land. I started to calculate when I could expect to reach St. Thomas. With mild trade winds, I thought, I might arrive in time to celebrate my birthday on December 28th. But the wind, refusing to coöperate, took a holiday. In the meantime, I hunted for a can of rye bread I had stowed aft. I had brought it with me to eat shortly before reaching land, in preparation for a fuller, richer diet. It served its function, for later changes in my diet brought no intestinal complications.

The *Liberia* wove her way slowly to the west. The following morning I stood south of Nevis, which lay before me, bare of clouds; but as I watched, a fresh cloud cap collected over the three-thousand-foot crater and, within minutes, grew into a beautiful white wreath. As I sailed east of Nevis I saw all the more northerly islands; even the farthest, Saba, was visible in a murky spot of cloud. I sailed on slowly, passing St. Kitts, shrouded in mist. To the north, St. Eustatius, which had profited greatly during the American Revolution as a neutral free port, became clearer and clearer. Then I reached Saba, a crater rising sheer and sharp out of the water, and here I slept. It took me a whole day to get out of sight of this island; the trade winds were evidently uninterested in my idea of a birthday celebration in St. Thomas.

That night a West Indian sailing ship, bound for St. Kitts, sleeked by. Then, on my birthday, I came in sight of St. Croix, but St. Thomas lay still further to the north. When I was northeast of St. Croix, the wind changed to the northwest, driving away all hope of reaching the other island. It became apparent that a storm was brewing; wind clouds rushed high into the blue sky, and the waters darkened. I held course as far west as I could, thereby gaining a favorable position for entering the harbor of Christiansted, the largest town in St. Croix. During the night the wind stiffened, the sea roared, and at dawn I was happy to set my course for the harbor entrance. Slowly, slowly the deep, dark water changed to a shallow, light blue. Rocks gave the impression of lying only a few inches under the keel in the clear water, and I had to reassure myself by checking a special chart before I dared sail on. Before me the town straggled up the side of a hill; to starboard, a small island in the harbor grew bigger and bigger, sheltering in its lee some twenty yachts.

I sailed up to a wharf where Caribbean sailing vessels were loading and unloading, handed my line to the usual crowd of dockside loungers and stepped on shore.

5 THE LAST STORM

As I jumped ashore I was surrounded almost immediately by a large crowd. "Where do you come from?" they asked. "From Spain, from the Canary Islands, in sixty-five days," I answered. I was interrupted by the arrival of the harbor master, who asked me the same question. He found my answer hard to believe. "In *that?*" he asked incredulously, pointing at the *Liberia*. I explained to him that it had taken me nine weeks, but I could see that he still found it difficult to believe.

I now moored the canoe alongside a coastal freighter, changed into more suitable clothing and, identity papers in hand, set off for the harbor master's office. My knees were weak; the ground under my feet felt as liquid as the sea; I found I had difficulty walking a straight line. But, on the whole, I felt extremely well and, above all, jubilant at my safe arrival. A deep sense of well-being and achievement made it possible for me to ignore minor physical discomforts. The crowd that followed me asked several times if this were really my first landing, for I walked fast, carried myself well, even managing the two steps up to the harbor master's office. I was examined by the doctor there, who certified that my general physical condition was good and that I had arrived without edemata. I was surprised to find that I had lost only twelve pounds; I had counted on losing at least twenty. I attribute the smaller loss to the fresh fish I ate, which enabled me to stretch my canned rations.

That same evening I found myself strolling through town in the company of four new-found friends. In local fashion, we shuffled down the narrow streets to the rhythm of a marching steel band. I ate my first steak and, later, sat at a bar surrounded by elegantly clad tourists. And I found myself thinking, "How strange!" I had the feeling that I had been in those surroundings for weeks, whereas, for the past sixty-five days my life had been so different. I had been rolled and battered in a dugout canoe, utterly alone. My beard, a painful boil on my thigh and the many questions I had to answer were all that reminded me of my voyage.

I spent ten days in St. Croix, enjoying good food, good weather and human companionship again, while the *Liberia* rested at anchor in the harbor. Then, my brief vacation ended, I prepared for my departure; the sails were overhauled, the ropes spliced, and fresh onions stored on board. I had planned to slip quietly out of the harbor, but my kind and hospitable friends insisted on towing me out in full view of all the anchored boats in the harbor, so that my leave-taking was public and noisy. We were hailed by cheers and shouts of encouragement, by the banging of saucepans and blowing of whistles; not accustomed to such excitement, I was terribly embarrassed. I was further discomforted when, as we were pulling out of port, the old towline broke—my reward for being too penny-pinching. However, the damage was quickly repaired, and I was towed to one mile beyond the coral reef. Here, in a dead calm sea, the towing boat and I parted company. As I left, someone on board shouted across to me: "We will see you again soon!" This aspersion on my boat forced me to paddle for several hours, until I was sure that we would not meet again.

Yellow-brown sargasso weed floated on the surface. Though I had looked for it before I reached the Caribbean, it was not until I was south of Antigua that I first saw some. I had crossed the Atlantic just south of the Sea of Sargasso, so named from the Portuguese word for seaweed, *sargaço*. In that giant area, forgotten by the trade winds and by the Gulf Stream, which cuts across the ocean just north of it, the weed grows in profusion. Millions of years ago, when life was only starting on land, it grew there, and it is more likely to survive a catastrophe of the earth than we are. During these millions of years, an animal world of its own has developed in the sargasso weed, and I was eager to observe what I could of it. I fished some out of the water. My attention was caught by the many little blisters that grew on the weed, for I could see no animal life. I chewed some of the berries and found that they tasted disappointingly like very salty slices of potato. I took another bunch out of the water, shook it out over the deck and found what I expected; an evil-looking caricature of a fish and some tiny dirty-yellow shrimp. Their ancestors had had sufficient time to develop protective coloring and to mimic the grotesque shapes of the weed itself; it was the art of camouflage at its best.

Among those Caribbean islands I made my first acquaintance with the American booby birds. They wear a milk-chocolate brown coat with a white shirt, and they are found wherever there is water. They are coastal birds, bad swimmers but excellent divers, who shoot like arrows deep into the water to catch harmless, unsuspecting fish. I have been told that they sometimes miscalculate the depth of their dives and smash their heads on rocks below the surface. I was disappointed in their flight, which was solid and unimaginative, lacking the aerial gymnastics of the shearwaters. They are often the victims of the frigate birds, who swoop down on them and force them to regurgitate their catch.

Finally, a gentle breeze came up, and I was able to set the mainsail. The boom, swinging from side to side in the high swell, knocked my sunglasses into the water, but, for fear of losing the *Liberia*, I did not dare dive into the water after them. Two small sharks, barely the length of an arm, swam curiously in the soft wake water. Irritated at the loss of my sunglasses, which resulted in my having to look directly into the sun, I was ready to vent my annoyance on the sharks. "Come closer," I commanded. Obediently, one of them drew nearer—and there—I ran my iron directly into his muscular head. But the fish, not in the mood for my game, hit the water dizzily with its tail fin, as if to rise in the air, and thrashed itself free. As the other shark approached, I smashed my spear down on its head. It was knocked out, but as soon as I had hauled it on deck, it came to its senses and slipped into the water, where it paddled wildly around and then disappeared. I was horrified by what I had seen, but it bore out what I had heard of sharks' tenacity. They have been drawn out of the sea, disemboweled, returned to the sea, their intestines put on a fishing hook—and whom does one pull out of the water again? The very same animal, eagerly devouring its own intestines. This is no fishermen's yarn but an authoritative account by a leading ichthyologist.

My present goal was Cuidad Trujillo in the Dominican Republic, but that evening I was thwarted by a shift in wind from the east toward the southwest. Disappointed, I put out the sea anchor and lay down to sleep. By next morning the wind blew so vigorously from the southwest that I was forced to take down the mainsail. After I had shipped a great deal of water, I decided to change course and sail to the more northerly island of St. Thomas. Despite a rough sea, the next morning found me in front of the wooded mountains of St. Thomas. Above me, planes took off or landed every few minutes; to port I sailed past uninhabited rocky islands. The seaport of Charlotte Amalie lies deep in a bay, protected by several small islands. As I rounded one of them, using the western entrance to the harbor, I had my first view of red roofs climbing up the sheer sides of mountains, of white palaces, of mansions built of native stone, of gaily painted villas, surrounded by palm trees. Directly in front of me, a fishing smack docked at a quay. I slipped in beside it, berthing the canoe.

A palette of painted frame houses clustered around a small green-and-blue church facing the quay. Fishermen arrived and threw down their catch. The scene was picturesque, but I needed a less active anchorage. Finding none, I sailed on to the yacht basin in the eastern end of the large harbor. There I found friends from St. Croix, with whom I settled down to a long, chatty lunch. Most of them were yachtsmen who had sailed across the Atlantic in large yachts and could imagine, therefore, what it was to have been at sea for more than two months in a hollowed-out tree trunk. During a lively discussion of sailing and sailors, someone remarked accurately that "it takes a damn fool to sink a boat on the high seas." I was flattered when these knowledgeable yachtsmen made me an honorary member of the Virgin Islands Yacht Club; their yellow emblem with three red crowns, is one of the most valued souvenirs of my trip.

For a whole week the wind blew from the southwest. When it finally righted itself, I sailed off during the night, thereby avoiding another noisy send-off. Outside the harbor I met an east wind that carried me through the Virgin Passage toward the northwest. The United States has military bases here, which make it advisable to steer clear of the many little islands that dot the passage. With a perfect wind of fifteen miles an hour, I was able to sight the lights of San Juan by early evening. But there my progress ended, as the wind died. I was unable to enter the marvelous harbor. Instead, I lay down in the canoe, only poking my nose over the edge every few minutes to make sure no steamer came too close. The following morning the surface of the sea still looked as if it had been lacquered, so I made myself comfortable and slept. Awakened in the early afternoon by a gentle rippling, I found that the wind had come up from the northwest. This shift, which could be a portent of bad weather, made it advisable for me to sail through the ill-famed Mona Passage between Puerto Rico and the Dominican Republic as soon as possible. The present wind would blow me through.

That night the *Liberia* sailed westward toward the passage under jib alone. I think this was the third and last time during the whole voyage that the wind did me this rare favor. The canoe gained only a few miles, but that was not as important as the wonderful feeling that while I slept, the canoe worked for me. The next day I sailed around the northwest cape of Puerto Rico, which was brightly illuminated by an airport. Planes roared constantly over my head, and the lights of steamers passed to starboard.

That night the wind gods wished me well, pushing me

through the tricky Mona Passage with a speed of ten to fifteen miles an hour. As I sailed, I was haunted to starboard by the small rocky Desecheo Island; sometimes it loomed close by, only to vanish suddenly and mysteriously. This was the only night of the entire voyage that I was unable to sleep or rest at all.

At dawn, standing south of Monito Rock, I could make directly for the southern tip of the Dominican Republic. Gannets and tropic birds kept up a busy traffic between the two Spanish-speaking islands. The next night the *Liberia* entered a lagoon that lies between the island of Saona and the Dominican Republic; the waves flattened, the water changed from deep blue to cobalt and then to light blue cut by streaks of yellow-green and turquoise. There was so little depth here, that the rocks, coral sand and plants on the floor of the lagoon were clearly visible. A handful of huts stood at the entrance to the lagoon, and the inhabitants came out to gape in amazement at my strange craft. Several times my keel crunched on the sand, until I finally ran aground. At that point I was struck with the idea of getting a photo of my boat. I took my camera, stepped into knee-deep water and took the first picture of the *Liberia* under full sail.

As the darkness increased a cold north wind came up, forcing me to put on every available stitch of clothing. Lukewarm water splashed in my face. The next day the *Liberia* seesawed her way farther along a coast that gave the appearance of being untouched by human hand. The shrill cries of royal terns filled the air above me; frigate birds hovered patiently above them, waiting to snatch at their prey. Late that evening the lights of Cuidad Trujillo emerged behind a projecting cape. I threw out the sea anchor in the lee of the cape and lay down to sleep, only to awaken a few hours later to find I had drifted far toward the west. The rest of the night and the following morning was spent tacking toward the entrance of the harbor. An ugly pelican, perched on a telegraph pole, sent pitying glances at the *Liberia* as she sailed up the Ozama River.

A harbor boat came alongside and offered to tow me in. "Not till the wind dies," I answered. The wind died away almost immediately after my proud answer, and I was forced to paddle upstream. A uniformed official, standing on the dock, ordered me over, and with the help of a rowboat, I managed to put the canoe alongside. I had been warned by American yachtsmen that officials in the Dominican Republic were likely to be obstructive and rude; so I was not surprised when the customs official shouted at me. In a few minutes, his chief arrived—dark-blond and slim, resplendent in a blue uniform complete with gun and holster in genuine wild-west style—and ordered me to his headquarters, where I proceeded under heavy escort. I kept my temper with difficulty, telling him that I had come to see the famous peace exhibition of Cuidad Trujillo, but that I would prefer to sail off at once if I was to be treated with such rudeness. The atmosphere became friendlier after the arrival of an acquaintance of mine, who vouched for me. Then the press came to interview me; curiously, the newspaper men appeared more impressed by my prowess as a linguist than as a sailor. In the meantime, smartly uniformed customs officers had examined my *Liberia* from stem to stern, inspecting empty fruit crates, canvas bags, and food supplies, even glancing through my books. After they left, I made order in the chaos, rescuing my underwear from the bilge water where it had been dropped, and closing up the bags. I berthed the canoe upstream, near a gaily painted shed marked, Club. From there I set out on a leisurely walk through the town, pausing in the harbor to admire a tree to which Columbus reputedly chained one of his caravels. In the center of town a guide steered me through the cathedral, Santa María, la Menor; he pointed out Columbus' bones, a point on which I was careful not to question him too closely, as I knew that Sevilla, in Spain, also claims the honor of their possession.

As I was eager to investigate the city, I hired someone to clean the boat, to remove the green algae beard and the long-necked barnacles from the *Liberia*. He also removed my watch, cans of food and clothing—a loss I noticed only later. After four mosquito-tormented nights, I left Cuidad Trujillo with a fresh land breeze, bound for Port-au-Prince, on the last lap of my adventure.

Outside the harbor the wind sprang up; in the east the weather cleared, and the lights of America's oldest city disappeared quickly. Terns squabbled in the air between the *Liberia* and the rocky coast; waves broke around me, their foam leaving light blue scars on the water. Although only under square sail, the canoe sailed dangerously fast, so that later that evening I put out the sea anchor and lay down to rest. In the middle of the night, disturbed by the shadows cast by the mountain tops in the moonlight, I hoisted the sail despite the wind. Late the next afternoon I sighted Cape Jacmel, a few miles to the north. Streaks of mist soon engulfed the coast, but every now and then mountain peaks pushed their heads through mist and cloud as if to observe the progress of the *Liberia*.

The next morning the sun swept away the fog, brutally exposing the barren mountain slopes. The wind died; malicious small waves climbed up the back of the high swell, which swept past the canoe in a great hurry, seemingly afraid the rocks of the Haitian coast might leave before it reached them. It ignored the wild, uncontrolled splashes of water from the smaller waves around it. Soon the last pitiful breeze died away, and only the majestic swell remained.

The *Liberia* seesawed so violently that I was afraid I might go overboard; my nerves were no match for this balancing game, so I took refuge in my hole after I had squirted the deck with water to cool it. At that moment I was standing approximately fifteen sea miles off the coast, to the south of Jacmel; my sea charts indicated a westward drift, but I found that the flood was actually shifting back toward the east. Optimistically, I peered out over the edge of my hide-away every few minutes to see if a breeze had come up. No sign. A long-armed garfish approached cautiously, showing off its blue coloring; triggerfish waddled around the canoe.

The next day brought another nerve-racking dead calm. Fishermen of Haiti, on these occasions, blow on a large sea shell—a request to the spirits of the wind to fill their sails. I was uneasy in the windless sea. The swell had kept its high, cunning character, sometimes appearing quite frankly to threaten me with its foam teeth. The swell I had encountered in the Atlantic had been gentler and more regular.

To the north of me, the mountains of Haiti disappeared into a blue mist, and the following morning brought decisive action from the weather. The first intimations of change came from the south, where a blue-black bank of clouds hung over the sea. As I watched it, it seemed to me that its aim was to devour me and the *Liberia*. I looked to the north and met another unfriendly aspect: mountains bathed in a dark sea of cloud that merged with the water. Only directly overhead did a blue sky still smile down on the *Liberia*. The two weather fronts approached, neither yielding to the other; it became evident that I had been chosen as arbitrator. On either side of me, clouds spouted forth torrential rain and the remaining blue sky above me turned hazy as small white, fluffy clouds were drawn southward by the wind. The storm, of which the high swell had been an omen, was on its way.

Nature could restrain herself no longer; to the west the sea bared sinister white teeth. The roar of the approaching storm rang in my ears. I readied the *Liberia* for her coming struggle by hanging the sea anchor astern, fastening the sails and pulling up the spray cover. Then the first breakers converged on the stern while the wind howled in the shrouds. The waves were short and hard and broke faster than in Atlantic seas. Within seconds the surface was a raging mass; crowns of waves raced furiously after one another; their foam turned to a high, thin spray. The *Liberia* became their toy. In their excitement these raging, wild, gigantic creatures threw themselves blindly onto the canoe, shook her, threw her from port to starboard, finally crashing and thundering onto the deck. I bailed whatever the sponge could absorb, but as I bailed, repeated waves brought more water with them. Then, slowly, the turmoil died down and normal storm waves took over. The sea is often more agitated when a storm starts than it is later. Now I was able to stop my frantic bailing; I even snatched a few minutes' sleep.

But I was wet through and through. Salt water burned the sensitive skin of my thighs. Although the roar of the storm continued, the whining and the roaring were too regular to disturb my sleep. I awoke instantly whenever breakers threw water into my hole; my system was so attuned that eye, ear, even stomach, reacted to these emergencies. Instinctively, I crept out of my hole into the dark night, groped for the sponge, sopped up water, squeezed it into a small pot and then emptied it overboard. A glance at my compass and I was ready for further sleep.

All that night, off Haiti, my floating log was flung wildly back and forth by the storm. In the morning the water foamed only a little under a clearing sky, but the seas were as tremendous as before. Off and on, giant, mountainous swells still broke over the *Liberia*, underlining the sinister, dangerous atmosphere around me. I could see the coast distinctly, even the high point of land behind Cape Jacmel. Tomorrow, I hoped, I would sail out of its range of vision. My hopes were dashed that night, when the sea roughened. I now had to bail again, a sure sign that I could not yet expect to sail on. I was trapped. There was nothing I could do to escape. Jacmel lay to the north in front of me; the wind blew as before; the seas roared and raged, thundering against the deck and hitting the port side with special force, as though they wanted to smash the canoe. But in this respect I had total confidence in the *Liberia*. I knew she was strong.

Later that night as if to taunt me, the *Liberia* suddenly developed a suicidal passion to lie athwart the waves. As I loosened the rudder, the boat vanished into a huge breaker, and it was only after a struggle that she emerged again. I adjusted the rudder so that the canoe lay once again with her stern into the wind. During this operation, water poured into my collar and down my body, leaving me drenched and miserable. I was again forced to bail, sometimes every five minutes, sometimes only every hour. The storm roared on; I lost count of the days I spent in this one spot, bailing, watching, hoping for a break in the weather.

Little by little, this delay so near my goal exhausted me. With these winds, sailing to Port-au-Prince was impossible. I would have to make Jacmel my goal. But even that lay tantalizing and unattainable ahead of me.

I checked on my sea anchors to find that the canvas of

one of them, although only three months old, had been torn out of its iron frame by the furious seas. My new, spare, sea anchors had been twisted so much by the force of the waves that the lines were almost frayed through. Only my plastic lines had held. I made myself a new sea anchor by fastening my last remaining steel cable onto a box cover, a blanket, some rusted iron spare parts and a full bottle. In fact, my last dispensable possessions were turned into a new anchor.

The fury of the Caribbean continued unabated, turning my impatience at my lack of progress into pity for my boat. Battered and rammed by storms, the *Liberia* was no longer a sailing boat; she had become a coffin with a lump of lead below and a sodden handkerchief for a sail above. Yet she had withstood the test better than I had. I was the captain, and her mistakes stemmed from my negligence or my ignorance. Then and there during that storm, it came to me just how reliable my canoe was, and, at the same time, how absolutely necessary it was for me to keep a constant check on the sea anchors—for on these rags, hanging astern, depended my boat and my life. Between curses and prayers, I trembled at the thought of any mistake I might make.

Implacable forces roared outside; breakers hammered and rapped loudly against the deck; the grim sea grumbled and broke, pushing and pulling at the boat. Vito Dumas, one of the most famous single-handed sailors of all time, faced seas like these when he sailed his yacht through some of the stormiest seas of our globe.

Dark clouds, engulfing the Cape of Jacmel, opened to give me an occasional glimpse of the shore; at intervals the sun pushed its rays through a barrier of black mist, but the raging of the sea continued. The canvas was torn to shreds by the rushing breakers, my skin sodden and tender from constant wetness. How long? how much longer? rang continually in my mind. I bailed. As I bailed, I was seduced by the idea of deserting the *Liberia*. I knew I could swim ashore with goggles and fins, but I could not bring myself to do it; I had to have patience and sit out the storm, it could not last forever. I sang my familiar songs until my throat was hoarse and dry. Outside, triggerfish cleaned the last of the barnacles off the *Liberia*, dolphins played happily among the infamous waves, the canoe rolled and shook and took on water.

On the ninth day after sighting Jacmel for the first time, the wind and the waves calmed, allowing me to set my sails for my last harbor. With the open ocean behind her, the *Liberia* entered the dirty tidal waters of the Gulf of Jacmel. To starboard, I sailed past a small fishing boat, whose crew—completely West African in appearance—gaped at me as though I were a ghost risen from the sea. Ahead of me Jacmel grew in terraces out of the water, like a Gothic fairy-tale town. I climbed onto my seat and peered at the shore line, trying to find an anchorage that would please the *Liberia*. I saw nothing. The drift pulled to the west, making it necessary for me to use the paddle as I cautiously approached the wharf. An audience had already collected there, watching me paddle desperately to avoid a last-minute shipwreck. With sail and paddle I made progress. A policeman, standing on the wharf, whistled at me, indicating that I should dock near him. Sail furled again, a few strokes of the paddle, a rope thrown to the crowd on the wharf—and the *Liberia* was moored.

I was unhappy with the anchorage: the high swell broke only a few yards behind the canoe, so that she staggered wildly from side to side, slamming the rudder against the wharf. I climbed ashore to negotiate for a better anchorage and was advised to take my possessions out of the boat and anchor her outside the harbor. In a few minutes, with the help of the crowd, my meager baggage had been transferred to the customs house. In contrast to his neighbor in the east, even the chief of customs lent a hand. The *Liberia* was soon bobbing in clear water, safely berthed, while I found shelter at the local hotel, the Excelsior. I took a bath, with water that came from gasoline drums on the roof. Then I had time, at last, to think over my past three months.

I thanked God that He had allowed me to make the voyage, bringing it to a safe end. My goal, Haiti, sister Negro republic of Liberia, was attained. My adventurous voyage had proved that primitive vessels, although unable to sail against the wind, can not only cross the ocean but can reach their goal. Was mine the first African canoe ever to touch American soil? I doubt it. But the *Liberia* is the first canoe ever to be sailed *single-handed* across the ocean, and I am certain from my reading she is the narrowest ever to have achieved it.

I was satisfied with the physical preparations I had made for my voyage: my food supplies were well chosen, my navigation satisfactory. From a sailor's point of view everything had turned out well. But as I examined my experiences of the past three months, I realized that physical preparedness was not enough; spiritually, emotionally, I had not been ready for my adventure. When danger struck, I lost heart too quickly and doubted the outcome; I had allowed fear to take hold. At the time of the disastrous loss of my rudder, I had even contemplated giving up the voyage entirely. Would it be possible, I wondered, to be so disciplined and in command of oneself that fears and doubts could not weaken one's resolve? For the time being I pushed the question aside, but I was determined to give it more and serious thought once I was home again.

During the last, wet, stormy days before Jacmel, ulcers had

formed on my body. Doctors at the local hospital obligingly gave me penicillin injections against these rebellious skin eruptions. This last trace of my discomfort soon disappeared.

A few days later, my possessions dry, my skin back to normal, I left the unpretentious charms of Jacmel for Port-au-Prince. At ten one morning I said good-bye to the *Liberia*, which was to follow me some days later by an overland route, and took my seat beside the driver of the bus that was to take me across the mountains of Haiti.

During the two months I spent in Haiti, enjoying new sights, sounds and a most hospitable people, I found much that reminded me of Liberia. At that time, 1955, Haiti and Liberia were the only two black republics in the world. Now there is another, Ghana, and tomorrow there will be many more, which will have an easier time achieving stability and prosperity than the early pioneers of Negro independence. Liberia and Haiti suffered from a prejudiced world, but survived their many years of isolation and struggle. Liberia, before World War II a primitive backwater, now has a stable economy and is justifiably proud of its progress. The history of a country is always one of ebb and flow; the strength and potentialities of the black people were underestimated by the rest of the world because it was unaware of the disastrous effects which a tropical climate and a strong belief in magic have on man's progress. Today, with the help of modern medicine and education, Africans have proved to us that they can handle pen and scalpel with as much skill as they have handled spears. The mere fact that Liberia and Haiti survived the rain forest, the mountains, the malaria-carrying mosquito—all so hostile to progress—gives proof of skill and faith without parallel in history.

cond Voyage

6 RESOLUTIONS AND PREPARATIONS

I returned to Hamburg in April, 1956, already contemplating a second voyage across the ocean. While sailing the dugout canoe, I had often thought back on the years when I had sailed a rubberized canvas folding boat at sea, and I found myself slipping into daydreams of a folding-boat crossing of the Atlantic. But it seemed an impossible dream and nothing more. It would be too difficult, too dangerous, too much would be demanded of me—my experience in the canoe told me that.

It was not until I learned something of voodoo in Haiti that I began to give really serious consideration to my new plan. Through voodoo I learned that one can, by deep concentration, a kind of self-hypnosis, change one's fundamental attitude toward a problem, that, ultimately through voodoo, one can rid oneself of fears and doubts. "Impossible is not Haitian," runs the motto of the newspaper in Jacmel, owned by my friend M. Brun, and this motto I took for my own.

Therefore, on my return to Hamburg I read everything that could teach me how to develop self-mastery and conquer my anxieties, for I knew that self-doubt and hesitation were my worst enemies in danger. My first step was prayer, the invisible weapon of man, which brings him healing power and relaxation, recovery and renewed energy. True prayer penetrates the unconscious, bringing peace to the individual and thereby helping him to overcome disturbing traits in his character. Without self-mastery, achieved through prayer and through concentration, I knew my voyage would fail.

The problem I had to tackle, first and foremost, was sleep; my experiences on the first crossing proved to me that lack of sleep leads to delirium and hallucinations and from there to deadly danger. If I wanted this second voyage to succeed, I had to train myself to sleep in short intervals, to exist for a whole week without regular, long stretches of rest. I remembered the system—a form of self-hypnosis—advocated by the American, J. H. Schultz, which he called autogenesis training, whereby one concentrates to such a point of relaxation that the environment is forgotten and the self is found. I had made good use of this method before when I had trained myself to snatch a few minutes' sleep riding home from work. So now I began to accustom myself to short intervals of sleep which would enable me to renew my physical and emotional strength when at sea.

Of major importance in my preparations was the need to create within myself the assurance of success. I had to rid myself of all traces of fear and self-doubt, so for three months I concentrated on the phrases: "I will succeed" and "I will make it." I hoped to make these thoughts second nature. At the end of the three months my whole being was permeated by a strong conviction that I would succeed; that, no matter what happened, I would survive my trip. It was only then that I decided definitely to carry through my plan.

I only told my closest friends of my intentions. I bought the standard Aerius folding boat made by the German Klepper Company, planning to make the necessary changes in it myself. The plywood framework of this boat can be folded like the frame of an army cot, its hull is made of strong five-ply rubber and canvas, the deck of clear, royal blue canvas. Air tubes, built into the hull at the gunwales, give the boat added buoyancy. Its length is seventeen feet, one inch, and its width only thirty-six inches. It weighed approximately fifty-nine pounds. The little harbor town of Las Palmas in the Canary Islands was to be my point of departure again. I chose it because of my many friends there, because of its favorable weather conditions and to avoid a dangerous and time-consuming coastal trip. When I arrived I found that nothing had changed since my previous trip; the employees of the hotel were happy to see their *navegante solitario* back. I was given the same room and slept on the same lumpy mattress. It was comforting to be in such familiar surroundings; I knew the shoe-shine boy in the Santa Catalina Park, I knew where to find the best pastry in town, and the members of the yacht club were all my good friends. A newsreel with scenes of my first trip had been shown a month before, so that I was recognized and spoken to on the street more often than in Hamburg.

My boat had not yet arrived, so I spent my time reading the case histories of castaways to gain insight into their psychological problems. I came upon the report of a French ship's doctor, Jean Baptiste Henri Savigny, who with other castaways had spent thirteen days at sea after his ship sank in 1816. What happened during those thirteen days is hardly credible. Desperation and mass hysteria took over; suspicion and panic spread among the survivors, some committing suicide, others jumping overboard in a useless attempt to swim to safety. Murder, even cannibalism, gained the upper hand. Human excrement was eaten. Out of the one hundred and fifty men who survived the shipwreck only fifteen were left. Those who died, died because of a moral breakdown that could have been prevented by disciplined leadership.

The observations of another ship's doctor confirmed the importance of morale; the castaways of his ship, the *Ville de Sainte-Nazaire* which went down in 1896, were adrift for seven days in an open boat. By the second day delirium and hysteria had taken hold of every one of the survivors. And one year later, when the *Vaillant* sank, only one quarter of those saved survived the six days in a lifeboat; here again death was due to a breakdown in discipline, for we know now that castaways can survive for nine days in the temperate zones, without food and water. Why did so many die unnecessarily? It was due, I believe, to their inability to adapt themselves to new conditions; the crumbling of morale and discipline was followed by physical calamity. One can cite many instances from the last war of castaways who could not survive the trauma of shipwreck or who, through ignorance, handled their food supply with suicidal stupidity. We see what men can endure as castaways when we consider the voyage of Captain Bligh of the *Bounty;* with eighteen men and a mere handful of food he sailed 3,600 miles in an open lifeboat under the tropical sun. His knowledge and his tremendous discipline brought the boat safely to Timor, after forty-two days sailing across the Pacific.

As I read the histories of castaways and shipwrecks, I became more convinced of the importance of morale, discipline and calm to the single-handed sailor. Friends who have sailed the Atlantic in yachts have told me of finding encouragement and peace in prayer during moments of danger. I, too, prayed, but I felt I should do more to prepare for my second voyage; I had to drill my whole being—conscious and unconscious—to accept my plan. So I continued what I had begun in Hamburg. I repeated to myself, "I will make it," and I added new phrases, "Never give up, keep going west," and "Don't take any assistance." Gradually I felt my doubts and fears disappearing, yielding instead to a really positive optimism and confidence.

I congratulated myself on having chosen a folding boat, or foldboat as they are often called, for now I would be able to relive exactly the feelings of a lonely castaway; I would share his suffering, his hope and his despair. I would know his thoughts during the long nights at sea. I would, in fact, have to contend with even greater discomfort than a person afloat in the life raft of a plane or a ship's lifeboat. By suffering to the utmost from the elements, I could test the durability of the human machine, and in a cockleshell like mine I would learn much that we need to know about survival at sea.

Experience had already proved to me that in challenging the sea I had picked an implacable adversary. But it is in the nature of man to better his own achievement; it is normal and healthy to strive continually for new records. Each newly established record, after all, makes a positive contribution by setting the limits of human achievement. Thus the athlete who has run the one hundred yards in 10.1 seconds is only satisfied when he has run the distance in 10 seconds. But it is a fact well-known to all good athletes that only the man whose past performance justifies it should try to break new ground. What is true for the athlete is also true for the sailor; I was able to challenge the Atlantic in a foldboat because of the experience and knowledge I had gained during my first voyage in a dugout canoe.

At last my foldboat arrived in Las Palmas and I was able to make my physical preparations. I christened her the *Liberia III*. My old friend, the sailmaker, sewed me two square sails, reinforced the stern by sewing canvas over it and brought the canvas on the starboard side, from which I could expect most of the winds and waves, as far forward as the foremast. The mast was reinforced by two backstays, and I made a mizzenmast by putting a paddle in a wooden socket-stick and slicing off one third of the blade.

President Tubman of Liberia stopped in Las Palmas on his way to Europe. He and his fellow-countrymen had shown a great deal of interest in the ocean crossing of a canoe of the Kru tribe, and the *Liberia II* is now in possession of President Tubman in Monrovia. For my second voyage he wrote a message of good luck in my logbook.

During my final preparations I continued my self-hypnosis. The last weeks before departure I fell into a mood of complete self-confidence. I had a feeling of cosmic security and protection and the certainty that my voyage would succeed.

7 AN IMPOSSIBLE VOYAGE

Departure: October 20, 1956

"Hey, Hannes." A friend's voice, hushed so as not to disturb the early morning quiet of Las Palmas, woke me.

"Coming, coming," I answered, as loudly as I dared.

I had spent the last night on the *Tangaroa,* a double canoe owned by my friends, Jim, Ruth and Jutta. As I awoke slowly, I heard footsteps from above, and Ruth's voice came down to me questioningly:

"Are you awake, Hannes?"

"Beginning to be," I answered and forced myself out of my bunk and onto the deck.

My friends were already up, waiting for me in the dawn. I expected to sail in about an hour. In the last days I had exuded self-confidence; yet, despite my techniques of self-hypnosis, I felt tension. I could not avoid an inner questioning: "Am I really doing this? Am I really crossing the Atlantic in a small rubber folding boat?" I countered these questions and calmed my nerves by repeating, "You'll make it, you will make it. Stop worrying."

"You act as though you were about to cross the harbor," my friends teased me; so perhaps my tension did not show.

A few last-minute touches were still required. The under water spear gun had to be fastened to the starboard side; Jim tied plastic material around the mast to keep it dry; Jutta sewed my spray cover; and Ruth handed me last-minute items from the *Tangaroa*. She also prepared a traveler's breakfast for me, fried eggs swimming in butter to give me a last boost of energy this side of the Atlantic. These three friends, convinced of my success, did everything to help and encourage me. Jim had advised me—as had friends in Germany—to use an outrigger when I was undecided between that and air tubes around the water line. I finally decided on an outrigger made out of half of an inner tube, sealed at both ends.

Then the sun came up; everything appeared to be taken care of, although I knew that I could spend another day in the harbor making changes in this and that, altering the trim of the boat in one way or another. But I knew I had to draw the line somewhere. This was it, now. The moment to leave had come. My friends tactfully spared me the painful emotions of a leave-taking; they climbed into their small dinghy and rowed over to the Santa Catalina wharf, a good spot for last good-byes. I paddled the *Liberia* through hundreds of fishing smacks over to a lobster crate on which stood two other friends. "You'll make it," they shouted.

I set the square sail forward and paddled slowly from the fishing harbor, passing the Santa Catalina wharf, where a handful of curious people stood watching me. Jim, Ruth and Jutta, motionless, were among them. I could well imagine what was going on in their hearts and minds, I was far from cheerful myself. They had their little dog with them, whom they had saved from drowning, and even he seemed nervous and anxious. Few people knew my exact destination, but many must have had an inkling. They remembered that before my last voyage I had told everyone that I planned to sail down the African coast, and instead I had gone across the Atlantic. This time, when I first arrived at the yacht club, my friends asked, "Hannes, are you sailing down the coast again?"

"Claro"—but, of course—I answered, only to be openly laughed at.

In the harbor basin I hoisted the gaff sail. Trade winds had blown strong for the past few days, but in the protected harbor I could not feel them. The clouds above me sailed northwest. It was now nine in the morning. I could hear cars honking from the shore and the whistle of steamers from the water. To hasten my progress, I took up the paddle again. It is an old rule for motorless craft to sail out of sight of land as quickly as possible. As I left the protected harbor behind me, the first gusts of wind flowed over my rubber and canvas craft, the swell heightened and the first waves wet the canvas. I passed Las Palmas on the left, in the shadow of the old cathedral spire.

Then from behind me I heard the sound of an approaching engine. In the high swell I made out a white object. "Aha! they are probably looking for me." A moment later I recognized the local pilot boat, and my hope of leaving without official obstruction was dampened. They were definitely heading for me. I left the sail up, but the boat came closer, and the men on it waved me back toward them. I ignored them. They came up alongside, the pilot shouting, "The harbor master wants to see you."

"Why?"

"I don't know. All I know is you're supposed to come back."

"I am sailing to Maspalomas, and for that I do not need anyone's permission."

Maspalomas is a beach to the south of the island, and it was true that I had thought of going ashore there to make some necessary changes in the trim of the boat. I sailed on. The pilot boat came up once more, this time on the port side, where I had attached my outrigger with a paddle. They drove right over it, breaking the paddle, and bringing me close to capsizing. I immediately took down the sail, trembling with rage at their carelessness. They threw a line to me. Furiously, I threw it back at them. The paddle blade was broken and would have to be fixed, so in any case I would have to stop somewhere for repairs.

"I'm going to paddle back," I shouted at them and they left.

I felt limp, tired and depleted. My paddling was too weak to get me back to the harbor. The pilot boat disappeared while I was still trying to paddle against the wind. I was only about three miles from the harbor, but somehow I didn't have the strength to get there. Perhaps it was the impact of my first disappointment, my first setback, and the sudden realization that I might not be allowed to make my voyage. No! I would not allow that to happen. My plans and preparations wasted because of a harbor master's whim? My savings, my hard-earned money thrown away? A voice inside me repeated, "I'll make it, I'll make it," and on a quick decision I turned around, hoisted the sails and set off again with a fine wind.

The trade winds freshened up, combers sprayed and splashed me. Nearly all the seas swept over the entire length of the boat, so that it became clear that she was too deep in the water. But as yet I had no time to take care of the problem. First I had to get the feel of the sea again and find out how to handle the loaded boat. The pointed bow lacked buoyancy, plowing the water to such a depth that at times the seas came to the foremast, while aft they ran to the mizzenmast.

I felt a little queasy. That morning Jim had reminded me to take pills against seasickness, as I do at the beginning of every voyage. Although the foldboat with its outrigger did not roll much in comparison with the dugout canoe, the motion bothered me, and I was glad I had taken the pills.

As I left the protection of the island, in whose lee lies the harbor of Las Palmas, the winds blew more strongly. Soon I took down the one-and-one-half-square-yard sail from the foremast and sailed with the three-quarter-square-yard sail of the mizzenmast. The aft mast was a paddle that sat on the aft washboard. I had arranged the steering as on the *Liberia II*, cables ran from the rudder to where I sat and I could control them with my hands or feet. Also, I had arranged the boat so that I could sit relaxed, leaning back against the aft board, with my knees straight out in front of me while I steered with my feet. I could also sit closer to the front with bent knees, still using my feet for steering. It was only later in the voyage, after I had eaten some of my canned food and the boat was roomier, that I could stretch out completely and then only on windless days when I did not have to steer. My friends had overestimated the speed of the *Liberia;* the loaded boat did not make more than three and a half knots. Now the little square sail on the mizzen was taking me south at two miles an hour. But as long as the trade winds blew I was satisfied.

At twilight I reached the beach of Maspalomas. As the wind blew directly toward the coast, the surf was too high for a landing. I remembered previous expeditions to the beach with my Spanish friends, when we had gone there to picnic and swim. Early that evening I rounded the southeastern tip of Gran Canaria. The sea was calmer, so I took the opportunity to check the outrigger and the paddle which had been cracked by the pilot boat. I took off the paddle and lashed the inner-tube outrigger to the deck, while I tried sailing for two hours without it. But it was too difficult. The boat rolled and my discomfort was acute. I had been rolled to last a lifetime during my first voyage. Ideally, I should have had an outrigger on both sides, but the boat was not strong enough for that. To keep her afloat in case she capsized, I had stored the inflated other half of the inner tube in the bow and several empty airtight containers in the stern.

October 21st

While I worked on the outrigger, the night passed quickly. In the darkness I heard an occasional clap from flying fish. I was worried by my spray cover, which I had made and waterproofed myself in Las Palmas, for water leaked through its two layers of canvas. Water washed continuously over the deck and the spray cover, leaking through to my knees despite my oilskin pants. At three that morning I wrote in the logbook, "The torture has started." Of course, no one knew better than I what awaited me, but even so I had forgotten that my skin was extremely sensitive to waterproofing ingredients. After these few hours of exposure it hurt badly; when I touched a spot on my body the whole surrounding area burned as though hot tar had been poured on it. It was impossible for me to change clothes in the heavy sea, and the pain became so great that I seriously considered turning back or trying to reach the African coast. But I had to keep on. I reminded myself of the phrases I had continually repeated in Las Palmas: "Keep going west; don't take assistance; never give up." I had hammered these words into my very innermost being back in the hotel when my heart and mind were calm. Now, in this moment of near-panic, I needed them. In my first anger at my skin trouble, my weakness came to the fore, but the often-repeated words kept me going.

During the day the wind came from all points of the compass; flat calm alternated with contrary winds and choppy seas till evening, when the old trade winds blew again. Butterflies fluttered over the water. I pulled a locust out of the sea, and wondered if some day someone might not do this to me.

Water continued leaking through the spray cover. With a little rubber syringe, the kind a doctor uses for cleaning ears, I drew it out. Then at last I found time to concentrate on a quiet meal. I had eaten nothing the day before except my hearty breakfast, but I had drunk twice my usual amount of liquid to prevent dehydration. I had deliberately avoided food for thirty-six hours, hoping to dull my senses by fasting and thereby to make the discomforts of the journey easier to bear. Now I drank my ration of unsweetened evaporated milk and ate several oranges. Suddenly I noticed that my leeboard had fallen off and was floating away. It would serve no purpose to worry over the loss; the ship's doctor of the *Liberia* had forbidden worry, for it could only sap my strength. I was better off in this one respect than most single-handed sailors: I had my own physician aboard. I knew myself well and knew that after the problem of thirst would come the problem of morale. I had to keep my sense of humor and a relaxed outlook; I had to remain cheerful, unconcerned and emotionally stable.

As the second night approached, I tried hard to concentrate on dozing off so that my strength would be renewed for the coming day. I had to learn the art of sleeping while sitting. This is an easily acquired skill, but I was forced to check on the course at the same time. I felt empty, drained of thought, my feet hardly able to control the course. Then, for a fraction of a second, I fell into a dream. A wave awoke me. In the morning I discovered that my head had sunk onto my chest in sleep several times. I realized that I was a simple man with simple needs, a tired creature who needed sleep urgently!

October 22nd

Yesterday I lost sight of land. Now I was really alone. For how long? I counted on seventy days. As I left the protection of the Canary Islands, wind and seas became heavier. In the morning a steamer passed two miles to port, unaware of the little *Liberia*. I sewed a plastic layer over the deck so that less water would come through the spray cover. Because my freeboard was too low, water washed over the boat. The wind was only twenty miles an hour, but even this was too much for the *Liberia*. I had to throw something overboard to lighten her. First went the quinces, lovely, sweet, Canary Island quinces, twenty-two pounds, quite unspoiled, floated off into the Atlantic Ocean. But the boat was still too heavy, so, very reluctantly, I got rid of another twenty-two pounds of canned food. The boat sailed better now, and I was less worried. The sea would replace what I had lost, as long as I had the means of catching sea life. I sailed on, carrying 154 pounds of food and drink for the seventy days that might still lie ahead. Despite my previous experience, I might have miscalculated. But it did not matter. What I had read of castaway reports from the last war convinced me that there is enough food in the sea to ensure survival.

I suffered intensely from the oversensitivity of my skin under the sun that day. It was so bad that I began to despair: "Jump overboard. Who cares about you? Who knows where you are? Not even at home do they know of your new voyage." These thoughts ran through my head, taking hold of my senses, until I drove them off by repeating, "Never give up. Keep going west."

My finger tips were swollen, the skin raw from handling wet articles and bailing the boat. In Las Palmas I had painstakingly developed calluses on the palms of my hands and a tan on my nose, but I had overlooked the need to toughen my finger tips.

As the sun set, the first curious tropic birds inspected the foldboat from above. These creatures all seemed bigger than I remembered from my first crossing. Was it perhaps because I felt so minute in my rubber boat?

And then again night came. The mizzen sail was set, demanding constant watchfulness from me to ensure that I stayed on course. Under the most favorable conditions a boat as flat as mine has trouble keeping on course with aft winds. I concentrated on putting myself into a kind of dozing coma in which my feet would still control the rudder, but no actively conscious thought would disturb my rest. On a clear night, when the stars were visible, I had no trouble steering with my feet, but with cloudy skies I had to use my flashlight to light up the compass.

My pulse that night had sunk to the slow rate of forty-two beats a minute; hunger, inactivity and my physically good condition played a part in keeping my pulse slow. My whole system was influenced, I am sure, by my total concentration on relaxation.

October 23rd

The weather improved, making it possible to dry out my drenched clothes. Above all, I wanted to expose my skin to the sun. My whole lower body felt as if someone had been sticking pins into it. I could hardly wait for the noonday sun to start my cure. At midday I put a paddle over the washboard, forced myself out of the opening in the spray cover and sat down cautiously on the palm of the paddle. I steered with my hands. Slowly, I peeled off my thin rubber jacket, oilskin pants, shorts, thick sweater and undershirt. Everything was soaked through. With a few clothespins I fastened my wardrobe to the shrouds of the mizzenmast. How superb to feel the sun warming and drying my skin! From then on the noon hour was devoted to health; it became my "hour of hygiene and preventive medicine." I sat high on my little seat, bailed water from the boat, dried my kapok cushion in the sun and made myself very comfortable. At the end of my hour I dusted my clothing with talcum powder, rubbed my body with a washcloth and dressed. Then I settled down again in the shadow of the mizzenmast.

The mild wind made a pattern of shingles across the surface; the *Liberia* rocked gently in the swell; and the ocean showed me its friendliest aspect. I felt peaceful.

I noticed a plank floating in the water. It was overgrown with barnacles, and two small crabs fell into the water as I picked it up.

Again the night put a dark veil over the sea. In small boats, nights at sea are disagreeable and uncomfortable. Fortunately my mizzen sail, which had been up since Las Palmas, protected me from the cold wind. My hope and ambition was to leave it up till I reached St. Thomas. Contrary winds, of course, could force me to take it down. I had chosen

St. Thomas as my final destination. I felt an obligation to the yacht club there, of which I was an honorary member, so my bow pointed straight to the Virgin Islands.

I was beginning to feel that dozing and emptying my mind of all conscious thought, that concentrating on nothing, would not, in the end, be a sufficient substitute for sleep.

October 24th

I passed an agonizing night, knowing there would be many others like it to follow. It was cold. When the sun finally shone I cut up a large seabag and sewed it aft over the spray cover. I was delighted with this accomplishment. The wind blew favorably from the east and my skin burned less. But my buttocks were uncomfortable; I could only hope that the condition would not worsen.

I found that I daydreamed a great deal. Girls appeared in my dreams, but I knew that in a few days they would be banished for the rest of the trip. A hungry man generally does not think too much about women.

On land I pray regularly; at sea I prayed for alertness and for comfort. I found that praying, which can be a sort of sinking away, a forgetfulness of the outside world, strengthened my morale. The night was clear; the moon outshone the stars, so that I could barely recognize the planets. Inside the boat everything was soaked. The trade winds, as they had done a few days before, came from all directions; one minute from the north, the next from the east and a minute later from the northeast. In a small boat wind irregularities are much more noticeable than on a yacht. Crosswaves slapped the foldboat, giving me the feeling that we were running athwart the sea. But the compass showed that it was the waves, not I, that had veered.

October 25th

On that morning I was pleasantly surprised to find a bottle of orange juice that Jim had hidden for me. Later on, as I took my noon position, I found a photograph of Jim and his two friends, Ruth and Jutta, stuck into my nautical almanac. He had written across the top, "Dear Hannes, keep going west. Your friendship meant a lot to us." And Ruth's message, "Don't worry, you'll succeed," strengthened my determination.

I knew I could rely on the accuracy of my noon position. I measured the longitude only approximately, from the difference between Greenwich and local time, but I had tested my chronometer, and I knew it was trustworthy.

A little later I discovered a locust, clinging to the block on top of the mizzenmast. I christened him Jim in honor of my friend and then worried about how to feed him. The butterflies that fluttered by the *Liberia* had never thought of taking refuge on her.

Once that night I fell into deep sleep, to be awakened by the flapping of the mizzen sail. In my sleep I had dreamed that a friend had taken me to the safety of a harbor where I could sleep without fear of capsizing. My dream was nothing but a rationalization of my weakness in falling asleep.

October 26th

My first concern that morning was for my locust friend, Jim. He was still alive, and a little later I photographed and took movies of him from every angle. The wind still blew from the east. My luck was holding at the beginning of this voyage. By now I was completely familiar with my outrigger and knew its reactions in all weather conditions. I had fastened it to the port side, the side that is more on the lee during a crossing like mine than the starboard. The Polynesians carried their outrigger on the weather side but mine was not like theirs. I, at least, preferred it on the lee. I had had great difficulty setting the sails on the dugout, but now, with the stability given by the outrigger, it was simple.

In order to lighten the boat, I ate only from my supply of canned food for the first week. My only uncanned food was garlic and some oranges. I took garlic along because it keeps so well under the worst conditions; not even salt water spoils it; it is also a better aid to digestion than the onion, although it contains fewer vitamins. In the morning I drank a can of evaporated milk, in the evening a can of beer, together with my meal of beans, peas or carrots and a few slices of garlic. Milk and beer raise the energy level of a hungry man. I have sometimes thought that on steamship routes, where rain can supplement the water supply, lifeboats should also carry milk. It is true that water is the beginning and end of a castaway's existence, but milk is more than a liquid, it is a food. It helps control the factors in a hungry man's body chemistry that make for panic and delirium. The ability to stay calm and in control of a situation is of paramount importance to the castaway. Expressed in simple terms, one can say that hunger creates an imbalance in a man's metabolism which, as we know from diabetes, may cause moments of delirium. The alkali content of milk and beer, as well as their easily absorbed calories, can counteract this condition.

In spite of my knowledge of the dangers of dehydration and the care that I exercised to prevent it, I noticed its symptoms in the cracking and peeling of my skin folds. Because man's thirst does not reflect his need for liquid this is a constant danger.

October 27th

Jim disappeared from the top of the mast. I looked everywhere for him, but he was gone. A pity! The wind weakened. The night before I had slept a little. I managed it by changing the sails so that the wind was freer and I could handle the boat more easily. Then I lay down on my left side, with my knees bent and my feet still controlling the rudder, my head resting on the washboard. It was no luxurious bed, but the sleep strengthened me. I slept for half an hour on one side, sat up and then turned onto the other side. I felt I had found a solution to my sleep problem and was only sorry that it could not be done when the winds were stronger.

Last night Madeira petrels danced around the boat. Mediterranean shearwaters, which regularly accompanied me during the day, were seldom visible at night. I wondered why it was so difficult to catch sea birds in these zones. Castaways from other regions of the sea have told of catching quite a few. The castaways of a Dutch luxury steamer that was torpedoed in 1943, hold the record; they caught twenty-five birds in eighty-three days afloat below the equator. It is possible that sea birds prefer to rest on floating rather than sailing objects.

The sun burned with pitiless intensity on my mizzenmast. I sprinkled the sail with sea water and found its shade cooler after that. I took the noon position, then sat down on my paddle seat for my "hygiene hour." I was amused at the thought of a steamer suddenly seeing a man take a sun bath on the edge of a foldboat in mid-ocean.

Before my departure I had developed a method which aimed at disciplining the blood vessels of my buttocks. For three months I spent fifteen minutes a day relaxing and repeating to myself: "I am quiet, I am quiet, my body is relaxed, completely relaxed. My thighs and buttocks are warmer, much warmer. Pleasant, warm blood flows through my veins." This was no cure-all for my problem, which came from the fact that I had to sit for so long in wet clothes, but perhaps it helped my circulation a little. I found it difficult at first to do this on the boat; possibly I was disturbed by my new environment, and in heavy seas I did not try it at all.

Now I began to see small fish under the boat. I had painted her underside red, just as I had the dugout canoe. I found out that little fish were attracted by the shadow cast by the boat, hoping it would afford them protection from their enemies. I am not sure as yet whether large fish, like sharks, will avoid the color red. I had brought with me certain other preventive measures against sharks; first in the line of defense was a piece of shark meat to throw at them in case of necessity (after three days it stank so horribly that I had to throw it away). Secondly, having heard that sharks are frightened by metallic sounds, I had tied together some pieces of old iron to throw at them in case of attack. I knew for sure that sharks do not intentionally attack boats. Other big fish follow the same rule. Despite this knowledge, I was still worried; after all, the fish might feel that here was such a cockleshell of a boat they had nothing to fear from it.

October 28th

That morning, looking for something on the boat, I found two presents from my friends of the *Tangaroa;* a bottle of rum about the size of a finger and a copy of the *Bhagavad-Gita,* one of the world's great religious works.

The trade winds freshened. Far too much spray and far too many combers washed over the spray cover, but in a foldboat these are as impossible to avoid as dust on a motorcycle. I felt myself as safe and sure on the high seas in a foldboat as a bicyclist on the road. In fact, in some respects I was safer; I did not run the risk of being pushed into a ditch. I owed this feeling of assurance to my previous trips in a foldboat and to the knowledge that the air tubes built into the hull of the *Liberia* would keep her afloat if she capsized.

My noon latitude that day was the twenty-sixth degree north. As I was returning the sextant to its waterproof bag, a dolphin took the bait I had hung over the starboard side. My first fish of the voyage! I hauled it on board and killed it with a knife. First I drank the blood, then I ate the liver and roe, which actually tasted better than the flesh of the fish, as well as being richer in vitamins and minerals. Later I ate part of the meat, putting aside the remainder in the shade of the compass for the next day's meal. I was pleased at having saved a whole day's ration.

The seas had grown stronger, bending the wooden framework of the boat with each wave. Because the air tubes were not fully inflated, there was a creaking, groaning sound in the boat. I began to feel that my own bones were making these unhappy sounds. Yet the foldboat sailed better now than earlier; sometimes she ran before combers like a Hawaiian surfboard. The *Liberia III* had certain advantages over the *Liberia II;* I found it easier to set her sails, and in calm weather I could read because she rolled less. Now the trade winds blew with full power, and the whole ocean appeared to move toward the west.

Fortunately for me, the nights were quieter than the days. At night I skimmed through a sea of bioluminescence, which enchanted me with its beauty and variety and reassured me of the ocean's fertility.

October 29th

Around midnight, the wind, calming for an hour or so, allowed me a little sleep. Then it returned, from the northeast, blowing with a force of twenty-five miles an hour. The outrigger slipped, and in the dawn a big breaker pushed it completely out of place. I waited for more light before repairing the damage. Then I took down the sails, forced myself out of the spray cover, readjusted the course and stretched out over the outrigger. I was pushing it back into position when suddenly an enormous, steep green wall towered above me. It hovered briefly and then crashed over me and the boat. I gasped for air. Because I had all my weight on the outrigger, the boat did not capsize, but it was half swamped with water. I climbed back in during a few seconds of calm and sat down in water. I adjusted my course so that the bow soon pointed west again; then I hauled out my little pot and bailed. I got up the last of the water with the rubber syringe. One hour later the inside of the boat was drier, but I was still soaked through.

The sea roared and stormed with such violence I was unable to take my noon position. A school of dolphins passed by, but I lost sight of them quickly in the rough water. Later that morning I spotted a few albacores.

I reacted to my enforced inactivity like a schoolboy; I squirmed on my seat, I wriggled, I moved first one way, then another, changing position every few minutes.

For ten days the trade winds had blown steadily. Had I hitched a ride with them that would take me directly to St. Thomas? In that case I could expect to make the crossing in fifty-five days. That evening rain squalls hit the *Liberia*, and with the help of my mizzen sail I caught three quarts on a plastic layer of my spray cover. I drank one quart immediately and put the rest aside in an aluminum container.

October 30th

Eleven days at sea. Now the trade winds appeared exhausted. The night was calm. An Atlantic swell of more than twenty-four feet rolled under the boat, heaving the *Liberia* up to a peak, from which she slid gently down into the next valley. At noon I enjoyed a thorough "hygiene hour." I washed all my clothing in the sea, drying it on a quickly installed clothesline. My knees and thighs were covered with small pustulae. I opened them with a needle, removed the pus and let them dry out in the sun. I fished purple snails from the water. As I crushed them, they stained my fingers with the dye that was once used to color the togas of emperors and kings.

October 31st

I had my first night of flat calm. My cans still took up too much room for me to stretch out in sleep though. The rudder already had a free play of more than twenty degrees, considerably more than I expected from my previous experience at sea. From the southeast a few small breaths wafted over a lazy ocean. I noticed a great deal of plankton, but as yet I had seen no triggerfish or water striders. The wind veered to the south, and a rainfall brought me more drinking water.

November 1st

My thirteenth day at sea, and the winds, of course, were contrary. My watch stopped working, but it was no great misfortune for the chronometer would replace it. A bad squall, coming from the southeast, broke my boom, as I laughingly called the stick no thicker than a finger to which the sail was attached. I repaired it with ease, but the sails no longer fitted as exactly. Before the twenty-four hours of the thirteenth day had passed, I suffered an attack of stomach cramps. My relaxation exercises brought some relief.

That afternoon the contrary wind stiffened. I had no sail up, but I was still blown back toward the east. So as not to drift too rapidly off course, I took down the paddle that served as a mizzenmast. For the first time on this voyage I threw the sea anchor out over the stern. I was surprised at how well the boat held herself, hardly swinging around at all. I attributed it to a good job of trimming the boat.

November 2nd

For a whole day I lay to on the sea anchor. The boat shipped a great deal of water, since we were no longer sailing with the waves or before the sea. Big waves ran pitilessly over the deck. I bailed at least once every hour, until I was so tired of it that I pulled up the anchor, so that I was again before the wind. I now faced the direction I came from. The only advantage was less water in the *Liberia*.

Around noon, the wind calmed. The latitude showed me that in two days I had lost close to forty miles. But when I compared it to the three thousand I had to cover, forty did not overly disturb me.

A giant swell, indicative of a real storm, rolled down from the North Atlantic. I perched proudly on my washboard, enjoying my "hygiene hour" and watching a little dolphin, hunted by three larger brothers, take refuge in the shadow of my boat. The little fish, not much longer than my foot, waited quietly under the air tube, but his hunters spotted him. One attacked but the little fish escaped. Now I watched a life and death chase. They swam around the rudder, showing very little respect for my boat. Finally one succeeded in capturing his prey. It was hardly an evenly matched contest. The three older dolphin measured three and a half feet to the little fish's paltry inches. However, they must have enjoyed the chase, for they started beating the bottom of the boat with their tails. They approached, turned on their sides and smacked the keel. Although they in no way threatened the safety of the *Liberia*, I was indignant over their lack of respect. In a momentary rage I grabbed my grappling iron, stabbed and wounded the first, then I thrust at the second, but broke the aluminum point of the iron in the process. The third dolphin continued his cavorting until I drove my knife into his body. I could have caught them easily with my underwater gun, but they were really too large to bring aboard without damaging the boat.

November 3rd and 4th

The trade winds returned during the night. The sky was cloudy and later a roaring, blustering and hurling squall poured rain on the sea, ironing it flat with its force. I caught fresh water in my spray cover. I had collected more in the past sixteen days than I was able to throughout my first crossing.

The next day I spotted a squid of some thirty inches. Its red-brown coloring resembled that of the giant squid I saw the year before, and like it, this little fellow appeared to have no tentacles. They must have been related.

The wind whistled through the shrouds, and at night I dozed only from time to time. I was beginning to feel tired. During the night of the sixteenth day I had the strange impression that the ocean ran towards the east. I could not rid myself of this feeling for some time. I put my hand into the water, and, sure enough, it ran eastward. The clouds stood still, and I believed that my boat was being carried back to the African coast. I shone a flashlight on the compass. It showed west. I knew who I was; I knew where I wanted to go, but heaven help me, how was I to rid myself of this certainty that I was going backwards? Everyone has experienced the same kind of sensation, sitting in a stationary train, when the train beside him moves, and he has the illusion that he is moving. But this hallucination in a small boat was unexpected. Then a comber washed over the boat, sweeping away with it this mysterious sensation. With complete assurance I knew now I was sailing west.

I almost always kept the mizzen sail up. Sailors will wonder why, but it afforded me such fine protection against cold winds that I overlooked the less advantageous steering.

The wind was still fresh, so the night was unpleasant. But the sea birds enjoyed it, for it made their flight easier and freer. I saw several Manx shearwaters. There were more of them this year than last, and this time I had no trouble identifying them. Because of their skillful aerial acrobatics, I nicknamed them "the mad flyers." My good friends from the first crossing, the Madeira petrels and the Mediterranean shearwaters, joined me once again. I saw only isolated tropic birds. On my previous voyage I had sailed closer to the Cape Verde Islands, where they appear more frequently. A trained zoölogist, by looking at the sea birds and at fish, can tell his exact location on the high seas without recourse to the stars.

The rough sea had forced me to do without my daily "hygiene hour." Without a sun bath to dry the pustulae on my thighs, I had to wipe them off with a handkerchief. I used my upper thigh muscles less than any others and properly I should have massaged them daily. However, the erupting exanthema made massage too painful, even though I realized that without it the muscles might shrink. Between my two voyages I intentionally gained weight so that I would use up excess fat rather than muscle tissue at sea.

"How much weight will you lose?" Jim asked me in Las Palmas. "About ten pounds?"

I told him I expected to lose at least forty-five pounds, and he pointed out that I could afford it, whereas with his skinny frame he could never stand such a loss.

November 6th

Finally the wind abated; the first trade wind clouds appeared in the sky. I put out all my wet possessions to dry in the sun: books, cameras, sea charts and, of course, my clothes. I knew I had sufficient drinking water to avoid dehydration and could afford the luxury of a long exposure to the sun. I noted with interest that this slight exertion brought my pulse up to forty-eight beats per minute, whereas before it had been thirty-four.

So far nothing unforeseen had occurred. The voyage had gone as I expected, based on my previous experience, except for less stable weather. My natural optimism took over; I planned new voyages or daydreamed of a farm in the tropics, always a pet idea of mine. During the first two weeks a woman appeared in my dreams of the perfect life, but as the voyage continued I rejected her completely; I was even baking my own bread. Food—I thought mainly of food, mostly of sweets; like many northern Europeans my special favorite is a cake with whipped cream. There, in mid-Atlantic, food played the foremost role in my daydreams.

November 7th

The trade winds lay dormant; only rain clouds and soft gusts remained unchanged. As I bailed the boat, I discovered that one tin of milk had corroded. In the afternoon a swarm of triggerfish plundered the first barnacles off the boat bottom. It was an odd sensation to feel them under me, first snapping at the barnacles, then diving deeper and slapping the rudder with their tails. My body was so much a part of the boat that I had the sensation they were attacking me.

November 8th

To one who uses his eyes, life at sea offers endless variety. Very seldom is one day like the next, but my twentieth day differed not at all from the day before: weak winds from the northeast, the same birds and the same swarm of triggerfish rolled under the keel.

I am an optimist—I have to be—so I celebrated the end of the first third of the voyage. Celebrated with what? There was not much choice; I could either drink an extra can of milk or an extra beer. I chose milk. I had also taken with me a pound of honey for each week of the crossing. I had not touched it yet, but as soon as I remembered its presence on the boat I pulled it out and in minutes half a week's ration was gone. It was difficult to hold to the rations I had planned for. During the first week I was sometimes hungry, but after that, hunger left me and I suffered only from thirst. I think I took with me the least amount of food of any boat that has ever made the Atlantic crossing.

November 9th

As long as the nights were calm, I was always able to catnap and renew my strength. During these naps my feet controlled the rudder, although I sometimes found myself off course, still the foldboat managed to stay with the waves.

I saw my first water striders. It is not unusual to find them in the ocean far away from land, although these were a different species from those I had seen in the Gulf of Guinea. The first swarm of flying fish swam and leaped alongside the boat.

I was approaching the tropics and expected to be south of the tropic of Cancer that night. I made several attempts to spear some triggerfish with my knife but only succeeded in wounding them. In the evening the wind freshened. Shearwaters' activity increased, while petrels behaved with the abandon of children released from school. Heavy seas in the early evening forced me to sail with care, lest the boat broach.

November 10th

My entry into the tropics brought no visible change; wind gusts were still fierce, identical birds and fish accompanied me and the sea—the sea was, as ever, full of twists and surprises. The wind had beaten the surface, stirred it up, finally turning it into a boiling inferno. Trade winds drew their breath of life from squalls and gusts, which every now and again rushed over the water. My heart felt for my boat, which had to make its way over a sea as rough and as full of bumps as an ancient cobblestone road.

I had to replace the bulb of my flashlight that evening, but my boat was so well-organized that I was able to find it at once. Day or night, I knew exactly where to find every spare part.

November 11th

Sunday, and my thoughts turned to the coffee and cake being served at home. Familiar church bells rang in my ears. Warm air wafted toward me from the southwest. I had decided to indulge in a Sunday treat of canned carrots; when I opened the can I was overjoyed to find Danish meat balls.

My mistake came from the trademark on the outside, which looked like two crossed carrots. I had suffered horribly from boils during my first crossing, and thinking they might have come from a meat diet, I took no meat with me on this trip. I had also chosen to take beer, hoping its vitamin B content would help prevent furunculosis.

I fell into a flat calm later that day. With a quiet surface I was always made aware of the plankton "dust" that floats near the surface of the water. I caught a little pilot fish with seven dark rings around its body; a good start for an aquarium, but unfortunately not in my present situation. Triggerfish returned. I hung a thin nylon line from the paddle support of the outrigger. It barely touched the water, and the triggerfish were fascinated by this dangling object. Every one of them wanted to snap at it, giving me a fine opportunity to catch some. I tried spearing them with my knife, but only succeeded in inflicting wounds. My hand proved more agile, so with a quick scoop I caught one around the head. At first it stayed motionless, but after a few seconds, it grunted a little reproachfully, as if to say, "You broke the law of the sea. I am taboo as food." But it was mistaken. Its organs had a wonderful flavor. Under their fins, triggerfish have red meat which tastes like meat from mammals and which I preferred to their white meat. I ate the brain of the poor creature, although it was not much bigger than the tip of my little finger. When we eat protein such as fish our body needs additional liquid to excrete the salts and urea, but there is no available liquid in meat, unless it is pressed out. The eyes, the brain, the blood and the spinal fluids do supply us with some water, but it is not enough to carry on the process of excretion. Therefore a castaway should not eat protein if he is dehydrated, for it will aggravate his condition.

An abscess had formed on the base of the exanthema on my thighs. I punctured it with a small incision.

November 13th

I noticed a new variety of petrel, with feathers so black that at first glance it looked like a crow. But its flight was as graceful as that of other petrels. I identified it as Bulwer's petrel. Three tropic birds paid me a visit, and one tried vainly to perch on the mast. Every bird gave me pleasure; I admired and envied their way of handling the wind, their easy, free and unaffected flight, their playfulness with the sea.

During the night one of the rudder cables broke. I was able to replace it at once.

November 14th

For four days the wind blew from the west. It was not strong, but I knew it was delaying my progress considerably. I was no longer alone on the *Liberia*. A crab had made its home on the port side of the boat. At noon every day he came out of a little hole and warmed himself in the sun. But he was a cautious companion, scuttling for shelter whenever I bent over to have a closer look at him.

I caught another triggerfish with my hands; this fellow had little transparent long-tailed parasites on his fins. A pilot fish popped out as I slit his stomach. With fresh bait I caught a dolphin. I sent an arrow from my underwater gun through his body, but could not pull him up on deck until I had killed him with my knife. I could only haul in a fish as big as this one when he was dead; otherwise, he would damage the boat. This fellow had many dark spots on his skin that looked like bugs. They were parasites, and under them the fish's skin had retained its light color.

I threw more bait overboard and caught another dolphin. But then my attention was attracted by the behavior of the rest of the swarm, that suddenly collected on the port side. A shark! Only three yards away from the boat a shark waited, and halfway between the shark and the boat my poor victim dangled from the line, fighting for his life. It looked as though the shark would attack the fish, but he was evidently afraid of my strange craft and did not dare approach. So I pulled the dolphin into the boat, killing him outright with a knife thrust between the eyes. The shark, lolling behind the boat, had obviously been attracted by the struggles of my victim. It was a fully grown shark, about twelve feet long, easily twice the weight of my boat, contents and crew. As the shark seemed as timid as I, I was less nervous, although I was glad to see it turn away. I looked over my catch. In the stomach of one I found a remora, a sucking fish, and in the other fish only my bait. I ate their organs first, then drank their blood mixed with a little rain water.

When I curled up to nap in calm weather, I could feel the snaking movements of my boat right through the rubber. But I trusted to the strength of the material. At that point in the voyage my mood was one of utter confidence; no disaster could touch me. I asked myself, "Why should I concern myself with something that had not happened yet and that I could not change anyway?" Although the calm dampened my hopes, I still expected to arrive in St. Thomas by Christmas.

November 15th

A warm breath blew over the ocean from the west. The atmosphere felt sticky; unlike windier days the air was empty of birds. I assumed they were resting on the water, storing up strength. I ate some more of my catch. It stayed edible for twenty-four hours if I left it in the shade; thereafter it soured. Fish can be kept longer if it is sliced and dried in the sun, and no water is allowed to touch it.

My attention was attracted by the snorting of a single whale. Then I saw his body, a black patch in the high swell. Occasionally a Madeira petrel fluttered listlessly in the humidity. Only two tropic birds retained their energy, flying high above me. Then once again a loud snort sounded across the water, and I saw a huge spout shoot into the air. I heard the whale take several shallow breaths, then one loud, deep breath; with that he plunged into the depths, his tail fin erect in the water. There is a whale skeleton on display in the Oceanographic Museum in Monaco, into whose abdominal cage the *Liberia III*, with hoisted sail, could fit quite comfortably.

Slowly, slowly I paddled against the breath blowing against me from the west. The sea boasted a pretty shingled pattern; I did about one knot with my paddle. I tired quickly and rested often. Without warning, a shark appeared beside me. It was about nine feet long, the average size for a high-sea shark. It stared at me out of its round, pig eyes, so close I could have reached out and touched it. I found my camera and took pictures while the fish dawdled under and around the rubber canoe. I had time to get at my movie camera, because the shark continued circling around the air tubes, surrounded by a host of pilot fish. I was not particularly eager to have a shark near me, but I was not very sure of how to get rid of it. I watched closely until it came within reach; then I hit it on the head with my paddle. My action had no effect whatsoever. After a while, however, it swam off, leaving me convinced that it had had no bad intentions toward me. It behaved as a zoölogist would have expected.

My sea-anchor line was no thicker than a pencil, but it

never occurred to a fish to bite through it. They do not destroy for the sake of destroying, nor do they plan or reason. They never thought of capsizing the boat and then attacking me.

I was worried about my right knee, which was swollen just below the knee cap and very sensitive to the touch. A similar swelling in any other place would have caused me less concern, but the knee joint merits special care and attention.

For the first time since the beginning of the voyage, I trimmed the boat; I took twenty-five pounds of food from the front and stowed it behind my seat. As I still had a west wind, the bow now went better into the wind.

I drew an empty bottle, smelling of gasoline, out of the water. I cleaned it carefully and stored it away.

November 16th

Again the wind came from the west. For nine days I had now had contrary winds. The air was heavy, my body dehydrated, my saliva thick and sticky. My tongue stuck to the roof of my mouth. I was roused by seven Mediterranean shearwaters, flying on the starboard side. I had never heard them make such a loud noise before, and the harsh metallic sound of their wings frightened me. Seven in a group—I had never seen so many together.

The swelling on my knee had gone up. I injected a syringe, already filled with penicillin, into my right knee.

The terrible west wind frustrated my whole being. It tried my patience, made me nervous, ill-at-ease and irritable. I was ready to start an argument with myself. I knew these symptoms to be common to people in a state of starvation or to castaways, but knowledge did not help me. Again dreams of my farm and of pastry, topped by mountains of whipped cream, brought release from tension. Nothing else interested me just then. In my mind I was either working on my farm cleaning out the chicken coop, eating marzipan cake, planting trees or whipping cream. My God! What a Philistine I was at heart.

Another bottle, covered with big barnacles and crabs, floated by. It must have spent weeks in the water. I ate two crabs, thoroughly chewing the hard shell to protect my mucous membranes. A huge swarm of pilot fish had adopted the foldboat. They made daring side trips to the outrigger, behaving as nervously as though they were crossing an ocean.

In the afternoon I spotted a sea serpent; I heard a quiet snort on the port side from a comber. I looked and behold! the fabled sea serpent. The swell took me to the exact spot, and there I saw four black curves, the first curve had a black fin. After the next swell I saw three more fins. Four little whales or porpoises, swimming after each other, made an impressive sea serpent. Soon they disappeared behind the high swell, and with them went every sea serpent legend I had ever heard of or read.

November 17th

During the night the wind freshened. I dozed for only seconds. The weather did not look promising; a dark, ominous wall of clouds concentrated in the east. My eyes turned to it constantly. Thunder rumbled across the water; lightning zigzagged through the cloud bank. I was still at twenty-one degrees north and at a longitude of around thirty west. I had sailed more than one thousand miles from Las Palmas.

Darkness came too fast and too early for my taste; in a few minutes I sat in a pitch blackness, through which white foam caps glowed, ghostly and insubstantial. The wind strengthened. I had already taken in the foresail. Now it thundered and lightninged all around me. The wind was more powerful than ever before on this trip. I still felt confident but no longer as cheerful.

November 18th

I passed a cruel night. Every few seconds I had to beam the flashlight at the compass to check on my course. I experienced the darkness of the blind, the thunder of the gods. I was battered, cold, wet and exhausted. The morning found me as empty and lifeless as a doll. I looked at the heavy seas and was afraid they might devour me. It was only fear—the fundamental fear of death—that forced me to stay awake, to use every last possible reserve of energy. Tropical rains smashed down, pounding the boiling sea. It was a storm of major proportions. Threatening clouds obscured the sky. I peered behind the mizzen sail, hoping to find a small crack in the gray sky, but the blue of the sky seemed to have vanished forever, and even the sun no longer existed. The wind raged at more than forty miles an hour. As soon as the rains stopped, the seas reared up to a height of twenty-five to thirty feet; some of the waves that rushed under me seemed even twice that height. I was surprised but pleased to discover that my mizzen sail was just the right size for such a storm.

My average speed through this boiling sea and frequent rain squalls was close to three knots, a good speed for the *Liberia III*. Most of the breakers, when they came from aft, pushed me forward without washing over me. On the other hand, when they came from starboard or port, the *Liberia* disappeared into a mountain of water. She could not go with the waves. It was an amazing spectacle of nature: the beating rain, peppering and battering at the water, turned it from green to white. When the rain ceased, giant combers overlaid the surface with white froth and foam. I felt I was in a supernatural elevator that descended rapidly to an inferno at the bottom of the sea, to rise again, with incredible speed, into the sky. But the rubber boat wound her way through all danger. It was impossible for her to keep a straight course with rear winds and a mizzen sail. After each rise and fall I had to look out like a watchdog to keep on course. Cross waves washed over the boat, but her buoyancy brought her up every time. My shoulders were battered by mighty combers hitting from the rear, they broke and foamed over the deck, stopping thirty to sixty feet in front of me. Any nerve-racking backward glance showed me combers capable of knocking down a house. A forward glance, onto the backs of waves, was less discouraging and dangerous-seeming. The trick was to hold to the westward course! Waves could not destroy a rubber-and-canvas boat as long as it went with the wind. The waves coming from aft had less power to do damage as long as they could carry the boat on their backs. Waves broke under the boat, oblivious of the four hundred and forty pounds they carried. I had the feeling that I swam in a cigar-shaped life belt.

The spray cover took on a great deal of water in this raging hell. I bailed every hour. My hands were soon bleached to a snow-white from rain water, the calluses on them soaked and swollen. When I did little repair jobs, my hands were easily scratched, but later the cuts healed without complications.

Despite the storm I had time to give thought to my companions; Mediterranean shearwaters flashed high in the air, dashed down again, rushing at the wave crests and making me think they had never felt better in their lives. Madeira petrels danced on the waves, touched the surface lightly with small black claws and obviously enjoyed the weather. Two big dolphins jumped out of the water, then let themselves fall back in, bending their bodies sideways in a fashion that always astonished me. My little triggerfish were busy cleaning barnacles off the boat bottom. These creatures behaved as always, taking delight in the storm; only the man in the rubber boat was ill-at-ease and worried.

My rudder had a free play of about forty degrees. This, combined with the mizzen sail, made it difficult to hold to the western course. I thought about the advisability of throwing out the sea anchor. I absolutely had to sleep in the coming night; without sleep I could not survive the storm. But I could never wholly forget the great danger of capsizing if I did not stay constantly on the alert. Then, there was a strong possibility that I might lose the rudder in such high seas, for the line of the sea anchor, stretching behind the boat, could simply lift it out. It was a difficult decision, but finally I threw out the anchor, and in a few seconds the boat lay in a good position. I took in the mizzen sail, drew the spray cover up high, leaving only face and chest free. I never fastened the cover over my stomach, as some foldboat users do in mountain rivers; it would have given me a trapped feeling and made it difficult to free myself if I ever capsized.

I thought about the famous voyage of Franz Romer, another story of suffering at sea. In 1928 he sailed in a foldboat, especially made for him, from Lisbon to San Juan in Puerto Rico. He was the first to do it. He left Lisbon at the end of March, shortly afterward was thrown ashore by a storm in the south of Portugal. He sailed from there to the Canary Islands in eleven days and from there made the voyage, in fifty-eight days of unbelievable torture, to St. Thomas. In St. Thomas he had to be pulled out of his boat. During the roughest part of his trip he could sleep—or better catnap—only between crests of high waves, then he was forced to awaken to control his rudder. To me the story of Romer's voyage is the greatest of all sea stories.

Franz Romer in his foldboat and I in mine had to control our boats every minute. They were very narrow and had no keels, so without control they would have capsized. But I had the advantage of setting out a sea anchor, which Romer did not dare to do for fear of damaging the rubber of his boat or the rudder.

Only a man who knows these foldboats can imagine the torture that Romer must have withstood. Eight weeks of sitting or lying, almost always wet, with never a possibility of standing upright, surrounded only by waves and combers. Eight weeks condemned to a cockpit, to suffering, shouting and prayer. In 1928 Romer did not have the medicines of today against pustulae, ulcers and boils. He had only one asset; the patience of a yogi and the energy of a man possessed by an idea. Anyone who has ever tried sitting for twelve hours in a foldboat knows of the cramping pain, but he has no idea how this aches when one's body is covered with ulcers, which don't heal, which exude pus and burn like the fires of hell. Whenever Romer wanted to eat or adjust his sails, he had to open his spray cover and—splash—a bucket of water was flung in his face.

At the very end, Romer, a former ship's officer, made the fatal mistake. He left San Juan in September, the hurricane month of the Caribbean. He had been warned, but he was determined to go farther. Shortly after he left, a bad hurricane swept the area. Romer was never heard from again, nor did any part of his boat drift ashore.

Now I lay behind a sea anchor, more exposed to the combers than when I went under sail, because the boat did not go with the waves. Romer's spray cover was destroyed by a breaking comber. I had protected myself against a similar disaster by laying thin planks over the washboard and putting air cushions over them.

The night was black and stormy, but I felt an unusual exhilaration as I lay in the boat, listening to the roaring, howling and boiling that went on around me, while I sat inside, curled up for a short nap. I fell asleep in a few minutes. Then I awoke, all my instincts working, and without conscious thought, I bailed.

November 19th

The night was endless. Heavy dark storm clouds would not let the daylight through; only rain squalls, thunder, lightning and bailing kept me alert. As I expected and feared, I lost my rudder. The steering cables responded heavily to my feet; then suddenly they turned light to the touch, letting me know the rudder had left me. Fortunately I had brought a spare. I dug my chronometer out from its rubber bag and checked the time. Daylight was half an hour overdue.

It came at last, but the monotony of dark-gray skies between masses of heaving gray-green water remained unbroken. Without a rudder, the *Liberia* could not hold a course to the waves, so I shipped even more water. My hands looked bad. I peeled off the sodden calluses.

The rise and fall of the boat during the night had made it hard for me to rest. I was still very, very tired. I crouched and laid my head on the washboard, too frightened to sleep, and as I lay there I heard the spray cover whisper to me.

"Now come," it said, "be reasonable and lie down. Forget everything. Leave it. Let others do something. You don't have to do everything."

At first this conversation seemed perfectly normal, until I

remembered I was alone on board. Often, as I awoke, I looked around for my companion, not realizing at once that there was nobody else with me. My sense of reality had changed in an odd way. I spoke to myself, of course, and I talked to the sails and the outrigger, but the noises around me also belonged to human beings; the breaking sea snorted at me, whistled, called to me, shouted and breathed at me with the rage and fury of a living being.

I had to wait for the storm to lessen before putting on my spare rudder. The sky still threatened. From time to time it thundered and lightninged; tropical downpours and combers emptied buckets of water into the boat. I bailed mechanically and patiently. I gave up worry and thought. The storm had shown me that I could have confidence in my boat. It is only after a man has lived through a storm with his boat that he knows exactly what to expect.

November 20th

This was my thirty-second day at sea. The night was restless, but the wind had weakened. I decided to replace the rudder. With the wick of the rudder between my teeth and the blade tied to my right wrist, I slipped, fully dressed, into the water. The waves were still fifteen to twenty feet high, the temperature of the water lukewarm. With difficulty I swam to the stern. One moment I was under the stern, then the boat hit my head, and the next instant the stern was before me. So I took the stern firmly under my left arm, changed the rudder blade into my left hand, when suddenly a big wave tore it away. I cannot describe the shock; it was unimaginable. I reacted quickly, grabbed for the blade and luckily caught hold of the string attached to it. I could feel the sweat of delayed fright coming up inside me. The next attempt was successful; with my right hand I pushed in the wick, fastened it once more with string and crawled back into the boat. I tried to undress, but the seas, with winds still blowing at twenty-five miles an hour, were too high. Water had entered the boat while I worked on the rudder, dragging my cushions out of place. I righted them and sat down again. With the steering cables I again controlled the rudder. My new rudder was the standard size, whereas the one I had lost in the storm was only two-thirds as big. Back in Las Palmas I had decided that my heavily loaded boat did not need a whole blade and that a standard size would exert too much pull on the rubber stern. I pulled in the sea anchor, fastened it behind the mizzenmast and set the little square sail.

Now my legs shook from delayed shock. It would take time for my nerves to calm down. But I felt as though I had won a battle, and after the bailing was finished, I treated myself to an extra portion of milk. The seas still roared around me, the wind blew furiously across a slightly clearer sky. The sun was circled by a big rainbow, a sign of intense humidity in the atmosphere.

November 23rd

Little by little, the wind eased. I hoisted the gaff sail. But the weather did not improve. Squalls chased each other across the water, squeezing rain out of the clouds onto the suffering foldboat below. Every few minutes I looked behind the mizzen to be sure that no bad squall approached, which would force me to take down the big sail. A chart of my course would have looked like the movements of a snake. I was thirty degrees too far to starboard one minute; in the next, a wave pushed me thirty degrees to the south on the port side; but on the whole I managed to keep west.

Two great big dolphins had followed me for several days.

Whenever the wind abated, they beat their tails against the bottom of the boat and then swam slowly off. They could easily be caught, I thought, especially when their heads came above water. I knew they were too big for my boat, nevertheless I had a strong desire to get one. I loaded the underwater gun. My first shot was a bull's-eye. The arrow landed on the fish's skull. He jumped, he leaped into the air and lifted the outrigger as he did so. Quickly I jerked at the line and held an empty arrow in my hand. The hooks of the arrow had not taken hold in the hard skull. Perhaps it was better so!

As the sun went down—1,400 miles from the Canary Islands, I saw a butterfly flutter in the air. Only the trade winds could have brought it this far.

My knees had improved!

November 24th

Tropic birds, Mediterranean shearwaters, petrels and Manx shearwaters flew around me again. The wind was very tired, and all my sails were hoisted. Suddenly, the port-side backstay broke. I took down the gaff sail and, to my horror, saw a huge, dark box only a half mile away. A ship had come up on me without my hearing anything. I had no idea what they wanted. Had they stopped to pick me up? I waved my hands and signaled that I was all right. Evidently they missed my hand signals; I was too far away. They made a turn around the *Liberia*. Stubbornly, I kept the sail hoisted. As the freighter came port side for a second time, I could even distinguish faces on the bridge, crew and a few passengers following with interest my boats' maneuvers. I took pictures and shot film. A young officer jumped from the bridge to the main deck, megaphone in hand.

"Don't you want to come alongside?" he shouted.

"No, thank you," I answered, without giving myself time to think.

"Do you want food?" came the next question, and again I shouted, "No!"

He asked my name, and I asked him for the exact longitude. After giving orders to have the bridge reckon the position, he asked me where I came from.

"From Las Palmas. Thirty-six days at sea and with course to St. Thomas," I told him.

"Would you like me to announce your arrival at the yacht club there?" came the next question.

I told him, yes, and gave him my nationality. He gave me the exact position: 56.28 longitude, 20.16 latitude. The young officer found it hard to believe that I didn't need food, but at my insistence the freighter slowly got under way. The captain shouted a last "good luck" from the bridge; the engines started carefully so as not to endanger my fragile boat. Then the steamer, the *Blitar* from Rotterdam, took up her western course.

The meeting left me dizzy. My quick decision to refuse food was unnatural. Perhaps my mental discipline combined with my orders to myself in Las Palmas, "Don't take any assistance," had forced my out-of-hand refusal. I thought about their reactions on the freighter as they stumbled on my funny, small craft—which obviously could not hold enough food for a crossing—in mid-Atlantic. The ready offer of help from the captain made me happy. It showed me that men are never alone, that castaways can always hope and that there are men all over the world who help others.

My latitude was exact, my longitude was one degree too far west. Thus even in a foldboat one can take the latitude accurately. Although I knew that in a high swell and with heavy weather, it is not absolutely correct.

Tropical rains came down in torrents that night. I caught five quarts of fresh water, perhaps to compensate a little for my firm "no."

November 25th

A steamer passed at nine in the morning within three miles but with contrary course. In the high seas and winds of twenty-five miles an hour, they could not see me.

I wondered when the famous stable and sunny trade winds would start. Only at noon did the weather clear for a short time. Another butterfly lay on the water. Dolphins hunted flying fish, and tropic birds circled above the boat.

For the past two weeks some of my canned milk had been sour, but only a little had really spoiled. I discovered some cans with small holes in them and decided that the metal was too thin. The sour milk turned into an excellent aid to my bowels. On a small boat, a badly-functioning digestive system can become a real nuisance. Somehow or other, I had to solve the problem every five to seven days. When I found that my turned milk helped, I no longer worried about the taste.

November 26th

I counted my pulse rate at night; that night it was thirty-two, lower than the usual thirty-four. My body adapted itself more easily to my hazardous ordeal than did my mind. I was still convinced of a successful crossing, but sometimes I became restless and dissatisfied, cursing at the unstable weather conditions. On my thirty-eighth day, a typical stormy squall rushed to the west. In the northeast, explosive masses of dark clouds gathered.

November 29th

Yesterday was calm. I shot triggerfish, happ͏͏ of my food supply. During the night I hung fresh bait into the water; it soon gave off such a strong bioluminescent glow that I could read by its light. In contrast, the meat lying on my deck stayed dark and lifeless.

In the clear blue morning sky, little trade-wind clouds piled up into huge banks of fine weather clouds. The day was warm and windless. On the flat surface, water striders glided over the plankton "dust." A dark remora, the length of a finger, tried to get free passage under the outrigger. It was shy and nervous, darting to and fro between the boat and the outrigger as though denying its stowaway intentions. A little triggerfish showed interest in something on the deck; it turned on its side repeatedly as it swam alongside, looking up at me with curiosity. But it was as cautious of me as I was eager to catch it, and it never came too close for safety.

I was right in the middle of the Atlantic now.

At dusk a swarm of triggerfish swam over to me. I shot two, while the rest stole my barnacles. It was the first time they had eaten off the *Liberia* at dusk, and I thought it might be a school that felt at home with me and my boat.

November 30th

With a hard, metallic beating of their wings, a Manx shearwater and Bulwer's petrel passed near the boat. The weather was still calm. Now and then a gentle sigh passed over the water. I tried to catch it in my gaff sail and to achieve paddling speed. Again triggerfish came for a visit. I had the impression that they were less cautious in a swarm. I was able to catch three. I was curious to find out how one can squeeze fluid from a fish, so I cut up a few slices of meat, put them into a handkerchief and squeezed. I chewed on the handkerchief, but no liquid came out. I ruined a perfectly good handkerchief by the experiment. I tried again with a plastic bag, first poking a few holes into it with a sailing needle. I chewed it and in a few minutes had a mouthful of purée of fish but no fluid. I cut holes into the fish—again no liquid. Later I tried collecting water from the morning dew, but I found it was not possible on my boat.

The little remora finally found a spot under the outrigger. I wanted to tease it—a little—so I put my weight to starboard, the outrigger rose in the air, providing my little stowaway with an involuntary air bath. Evidently it did not care for the excursion into other spheres, for it detached itself from the outrigger when it hit water and went back under the boat.

Some hundred yards off to the north, a waterspout reached from the water to the dark cloud banks. If it hit the boat, it could sweep everything off.

December 1st

A tropic bird, a last farewell from the east, circled the mast. The dull, weary weather remained unchanged. It had the one advantage of curing my skin eruptions by permitting protracted "hygiene hours." I grabbed a triggerfish, but my stomach reacted against its flesh. I imagined that my breath smelled horribly of raw fish. But I was saving my food for worse times ahead. Early in the afternoon a ship passed by within two miles. The high swell made it impossible for me to determine whether she was a tanker or a freighter. I felt so well-prepared for what lay in front of me that I saw no reason for concern about passing ships. Sunday sailors who stop ships at sea and ask for position and food are like the man who pulls the emergency cord unnecessarily on a train.

The air felt thick enough to cut with a knife. I had hoped that a black cloud wall would bring a little freshness, but it brought only five drops of rain and fierce winds. But that evening, the rain did come, filling all my containers. I put the aft sail over my head and the rain water poured from it onto the plastic layer of the spray cover. From there, using a sponge or a syringe, I transferred it into bottles.

December 3rd

The sea roared again. A stiff breeze blew from the northeast. Occasional, dangerous squalls sped by; the air echoed to a rumbling and howling, reminiscent of a first-class storm. A tiny fish landed on deck. I ate it, but it was very fatty and tasted as though someone had injected rancid olive oil under its skin.

December 5th

Two bad nights. Several times waves struck the boat crosswise. I sailed the correct western course. Heavy seas coming from the beam nearly capsized me. I was saved by my outrigger. The whole foreship plowed the water in heavy seas, sometimes taking water over as far as the mast, though the narrow bow projected far out of the sea. It was good I had a mizzen sail; with a foresail the bow would have dug deeper into the water and I would have run the danger of somersaulting. Rain squall followed rain squall, while the wind, coming powerfully from the north, drove tons of water over the westwardly running *Liberia*. Another day's ration saved by eating fish.

December 6th

For five days the weather belonged to the wind. My body was constantly wet. Everything ached: knees, elbows, shoulders and—as one would expect—buttocks.

My attention was caught by three small articles floating on the water, one of which looked like a mouse trap.

December 7th

I was too exhausted to sail all night as I wanted to. Once again I heard voices speaking to me from various parts of the boat and I answered them.

"Where are you?" I asked the knife. "Come on, don't hide from me. I've got work for you to do."

To the outrigger, as a heavy cross sea tried to push it out of place: "Great! Show the sea what a half tire is worth. Don't, please, make a fool of me. And don't forget, you and I have to stick together in this. If you go, I go."

The imagination plays extraordinary tricks on tired ears; the breaking waves shouted, praising or cursing me. They whispered and talked to each other, to the boat, to the lonely sailor. It was clear I needed sleep. I put out the sea anchor and curled up to take cat naps.

In the morning the weather looked no friendlier; stormy squalls rushed over me, some wet, some dry. A northern wind took command and pushed me south. I tried to sail west, but it became dangerous. I was lucky I had not yet capsized; several times the outrigger plunged deep into the water. I touched wood and took the precaution of putting out the sea anchor so that I could sleep a little.

December 9th

The sea calmed; I shot at dolphins; one struggled free, another flew off with my arrow, which slipped from its nylon line. I had a replacement but decided not to try for big fish any more. I enjoyed a meal of triggerfish meat.

I developed an abscess on my right hand, where a triggerfish had bitten me; the wound healed very slowly, and my local lymph glands reacted by swelling up.

I ate my last piece of garlic. From now on raw meat would have to be eaten without flavoring. My thoughts ran in one consistent groove—food. I decided to stop in Phillipsburg on St. Martin Island—a Dutch West Indian possession—and shop there. I planned a menu down to the last detail: a big loaf of crusty French bread, slices of Swiss cheese and ham, sweet butter and a dessert of applesauce with cookies and chocolate candy. I would prefer, of course, a cake, but I was sure I would not find it in the tropics. Who knows? Perhaps my Christmas would be spent there.

December 11th

The weather did not change; dangerous clouds gathered in the north; the air was sticky and the wind sleepy. Sargasso weed floated past me but I did not allow myself any false hopes, although in these latitudes it is generally found near the Caribbean. My longitude measurements were rough but reliable, and I knew where I was. My noon position indicated a climb to the north. Menacing weather from the north interrupted my "hygiene hour." My practiced eye discerned a siege of bad weather ahead.

At three that afternoon it was still calm and flat. At six the clouds exploded.

The sea roared and tropical rains hammered at the water.

I felt like giving up on the trade winds. Should I sail southward? In a vile humor, I threw over the sea anchor. A nylon line, no thicker than a pencil, held it to the boat. A shark's bite and the line would be finished. But I knew this was unlikely. I remembered with amusement the statement I had read in an old book in St. Croix about a shark's taste buds. The author, stating that in the Caribbean sharks attack a Frenchman rather than an Englishman, gave as his explanation the fact that the fish were repelled by the latter's meat diet. I was sure that a fish with such epicurean tastes would never be interested in a nylon line.

December 13th

The trade winds blew at thirty miles an hour. Feelings of discouragement and disappointment took hold of me, and I found myself wishing for a taste of the fine favorable winds I had had during the last three weeks of my previous trip. I comforted myself with the knowledge that these ugly winds would have to stop sometime.

Twenty dolphins, flashing blue and green in the water, gathered around the *Liberia*. I had no time to watch them. My rudder needed my full attention; again it had too much play, and I had to concentrate to stay on course. I was surprised to see so many dolphins at once.

I avoided backward glances, for they showed me terrifying seas and towering breakers heading straight for my frail *Liberia*. Once I narrowly missed disaster. An enormous breaker, coming from the rear, left me gasping for air as it poured water onto the boat. We were taken thirty to forty-five feet high into the air and then flung down with a hard bump. It was my first experience with such violence, and I had no desire to repeat it.

To what did I owe the fact that I was still afloat? Luck?

Was it only luck that I still lived? I refused to answer my own question. I knew I was well prepared, well trained, and beyond that, I would not analyze my situation. But I had to admit that these giant breakers caused me concern. They did not come often: in a span of twenty-four hours they might pass only once, perhaps twice, with full strength—but the mere prospect unnerved me.

In the dusk I spotted a red light, a little later a green light and then both together. At first I could make neither head nor tail of them, when suddenly I realized that a steamer was bearing straight down on me. In a panic I flashed my flashlight on and off against the mizzen sail, put the paddle beside me ready for use. The ship gave the impression of coming at me head on. I watched with extreme relief as she passed about fifty yards away. I noticed that she rolled even more than I in the heavy seas.

December 14th

I sailed all night through. I had no recollection or feeling that I had ever slept. I knew only that I was tired; terribly tired! Often during the day my eyes closed and my mind wandered. A tropic bird from the western Atlantic, approaching that morning, gave me comfort. The first American to greet me on the voyage. I knew, of course, that they fly far from land, but still I welcomed him and cheered up. The wind blew from thirty to forty miles an hour. I forgot the beautiful white bird. I thought solely of my discomfort and my fears. I felt so small, so insignificant and so helpless in these powerful seas. The unusual name that a Fanti fisherman in Liberia had given his canoe repeated itself endlessly in my mind. "Who are You, Seapower?" My God, Fanti fisherman, I thought, come and look at these waves, and you will feel as small as I. All at once a huge steamer loomed to port;

it had come up without my noticing it. What do they want, I asked myself, and waved at them.

"Everything is fine here," I shouted. Then I saw a man, megaphone in hand, calling to me.

"My dear Lindemann," rang across the water, "don't be a stubborn fool . . ." And the rest of his words drowned in the roar of the waves. The words were spoken in German, and the voice was familiar. And then I knew who he was. It was the voice of a newspaper man whom I had met when I returned from my last trip. I remembered him clearly because he had arrived before the others to interview me. He was a former ship's officer, but I could tell at the time from the manner in which he questioned me that he hated the sea. And why was it this voice that shouted at me from the steamer?

The ship made a circle around me, putting oil on her course. But I continued sailing, passing by the oil slicks. The small breakers lessened, and the surface appeared smoother; but the huge combers were not impressed by the oil and thundered on. Another squall swept the sea and kept my hands and feet occupied steering the boat in a westerly direction. The steamer, circling around, confused me. I forgot to head west. I took pictures of it, and then there it was, alongside once again. A young officer made a despairing gesture in my direction. Could he not help me? Smiling, I waved a "no" at him, but my smile was a parody, a horrible grimace. I had begun to realize that anything could happen to me in such stormy trade winds, that factors I had not reckoned with could overtake me. The ship veered off close to the boat, its waves mingling with wind waves and splashing over my deck. Then I found myself in its wake, forcing me to be on my guard to avoid the log line of the ship. On her stern I read, *Eaglesdale, London.* The meeting cheered me, for it was fine to know that people wanted to help even though I would not give them the opportunity. Perhaps I should have accepted. Give up after eight weeks, after fifty-five days at sea? I had to succeed by myself. I would come through all right. I was determined.

The German voice coming to me from an English steamer puzzled me. Had I really heard it? Could the man have resolved his hate-love for the sea by returning to it? I would have to write to the ship and find out who had spoken. (After my arrival in St. Thomas I wrote the captain of the *Eaglesdale*, who replied with a friendly note, congratulating me on having survived "such bad weather." But the German voice was a hallucination. My eyes had reacted correctly, but my ears had deceived me.)

As the ship left she put oil out to calm the seas and prolong my life.

December 15th

I passed a night of hell. Again I had no sleep. I was afraid to throw out the sea anchor because in these heavy seas its line would threaten the rudder. I knew I must not lose it, but I also knew I simply had to have sleep. I must not overestimate my energy; I had to be fresh enough to stay on course. On the other hand, I could not afford to sleep even for an instant in that boiling sea. The problem looked quite different here at sea than it had on shore. At sea I could only stay awake for four days and nights; on shore, with short cat naps, I had managed it for longer stretches. And now I felt my eyes closing. I dozed, I dreamed, I became the prey of imaginings and hallucinations. And then I put my last energy into staying alert. I began to sing. Slowly, I ground out a tune, only to find that something in my body cut off my voice. Then I counted, one . . . two . . . three . . . four, and suddenly I could not find the next number; it was lost; it was simply not there any more . . . I knew only one thing; the boat had to go into a garage; somewhere I had to shelter her and lie down beside her and sleep. . . .

The mizzen sail beat against my shoulders. A warning? I flashed the light onto the compass and found I was headed too far north.

I had the feeling that behind me stood a barn in whose lee the waves were flatter, while farther out on both sides, the sea still raged. As soon as I left the protected lee, masses of water washed over the deck. Ah yes! The barn would protect me . . . stay in its lee . . . where it was calm, cal—me ——r and then . . . water . . . I swam . . . what was it? The shock awakened me, I flashed the light onto the compass; too far south. I did not hear the breaker that swamped me, it was simply there. I bailed, I had to bail . . . I must bail. Why wasn't I doing it?

I was invited on a hunt. A Negro servant called for me. Lovely! I trusted him, he knew where we had to go. I sat comfortably in a kind of rickshaw. I saw big white lines ahead, and they worried me a little.

"Boy," I asked, "where are we going?"

"It is all right. We have to go through the surf," he answered, and as he spoke we plunged through. The deck was under water and came up again. I looked at the boy to the left. He wore black and snorted like a whale or horse, but he worked without talking back.

"Boy, where do you boys live?"

"In the west."

West! The word reminded me of something. I knew it, and then I remembered the compass. Again I was off course. I looked at the boy on the left, but he had gone. A black horse rode there now, pushing the boat. Horses know the way home. I could rely on a horse . . . satisfied, I relaxed—then suddenly I seemed awake, slowly and instinctively I came to myself. But who was I? No answer. What was my name? No name. What was happening? West, west—and no more stayed with me. Again I remembered the compass. The flashlight lit it up. Again, off course. Then a sound came into my consciousness, the sea still roared. I was cold, although sail and spray cover provided ample protection. Then I clearly heard the voice of Mephisto: "I do not see your water lies." I looked for my black boy and black horse at port side. I saw only black outrigger. It must be more than a lifeless object. It had to have spirit and soul.

During the morning a real storm with winds of forty miles an hour had developed. I looked with disbelief into the face of the waves. "Such waves cannot exist," I thought.

A little later I screamed and shouted, "I will get through. I will make it, I will make it."

As if to confirm my optimism, I saw a frigate bird, an American frigate bird, sailing high through the air. According to my calculations, I was still four hundred miles from the Caribbean islands; but frigates rarely fly more than one hundred and fifty from their land base, and I could have made a mistake. What a lovely mistake! It meant landfall within four days. This called for a double ration of food right away. I had starved myself enough. Now began the good life; every day I would eat double rations from now on. I would celebrate Christmas on land. Lucky, lucky man. When I threw out my sea anchor I did it with happiness and a sense of relief. My secret aim was to celebrate Christmas on shore. I bailed and dreamed of Christmas pudding. The storm was at its peak. The boat had trouble sticking with the waves. As she had lightened considerably, she might now be badly trimmed. I should have done something about inner ballast, but both water containers, provided for that purpose, had developed holes. So I forgot it.

I woke up and bailed, napped again. I sat, the spray cover drawn over my head. The time was nine in the evening.

All at once a huge wall rose on my right . . . nothing more . . . out . . . empty . . . dead? No, I gasped for breath, beat with hands and feet, and then they were free. I had capsized, was in the water. "I must reach the boat, the waves must not separate us," went through my head.

The hull stood high over the surface. It felt slippery. My mouth tasted of salt. At last I caught hold of the outrigger. The boat lay across the waves. I pushed her into the right direction. Would the storm ever stop? What could I do? I thought back to the time when my boat had capsized near Madeira, on one of my shorter previous foldboat voyages, and I remembered the difficulty I had had then in righting her. I found myself between outrigger and boat, with only my head out of water. The storm showed no sign of subsiding. The waves rumbled, roared and thundered as before, mercilessly. In the sea, my body felt bitterly cold. Then I climbed onto the hull, my right hand on the paddle to the outrigger, the left cramped to the edge of the boat. The wind hurtled over the hull, comber after comber washed over me. Still I was terribly cold. Only my head, protected by a woolen cap with a hood over it, stayed warm.

Was this the end? No! I would not allow it to be the end. I willed the *Liberia* to stay afloat. Would I sail the rest of the way to the islands, perched on the hull? Waves, warmer than the winds, broke over my back. I glanced at the stars. Orion was not even in the zenith, so it was not yet midnight. I knew I had to wait till daybreak to right the boat. I faced seven hours precariously balanced on the hull. The stiff winds chilled my body, I slipped back into the water. My body curled, and with cramped hands I held tight to the outrigger paddle. Every movement stirred up cold water between my skin and clothing. I forced myself to remain motionless. During the night the sea calmed, but big waves still made giant shadows. I felt sick. I vomited—I must have swallowed sea water. My hands clutched the paddle. How strong they were! To keep my body in the right direction, I had to tread water constantly. I froze.

I thought of home and of my parents; they knew nothing of this voyage. They could not imagine what had happened to their child. Self-pity engulfed me.

December 16th

It was midnight. Nothing had changed, except that I was even colder than before. Then one of my feet kicked against something. What was interested in my feet? Anxiety attacked me. I scrambled up onto the hull, my legs gripping the westward-pointing stern.

Two A.M. Orion descended in the west. I had slid back into the water, forced there by the extreme cold. The pitiless winds had reduced me to a shivering, chattering skeleton. I thought back to a voyage I had made in the Straits of Gibraltar one winter, when the wind blew so cold that my hands were unable to hold the paddle. The wind, that time, blew me into a harbor around midnight, and I had been offered a drink. That was what I needed now.

Suddenly I heard the sound of bells. They reminded me of church bells at home, the same bells that I had rung as a child. Did they ring now for my funeral? They must know, surely, that I could not die now, that I would get through? I was quiet, my muscles held on, instinctively, and demonstrated that deep in my subconscious there was still life. What did the heavy, brutal combers want of me? Didn't they realize that they could not touch me. That I was taboo. That I would survive, that I would make it, must make it. I had not lost my faith, nor had my subconscious or my instincts betrayed me.

At four A.M. Orion was about forty-five degrees west. I dozed on the hull. Once a voice invited me to go to a nearby farmhouse, to have a drink and sink deep into a feather bed.

"Where is the house?" I asked.

"Over there, in the west, behind the hill," came the answer.

Then I awoke. My sense of hearing had returned, I could feel the numbing cold again. I heard myself repeating aloud: "Don't give up, don't give up, you'll get through." I dared not sleep. Deep sleep meant certain death. I knew the sea devours everything, leaves no trace, draws even the dead downward. In the water again, I floated, dead, empty of feeling, at times delirious. But something survived, the lighthouse that guided me was my determination to succeed. As long as I had that, I lived. Sometimes, the lighthouse darkened—then there was nothing—only muscles—an animal without thoughts—all instinct, until the lighthouse suddenly lit up for me again. Loud and bright, it warned me not to give in, to keep on fighting. It shouted at me, "You will make it." Then I awoke, my senses returned, first hearing, then feeling, then speech. I heard death in my ears, sea and storm beat upon my body, salt stung my eyes, cold shook my bones. I was grateful for my lighthouse, it made me a slave who was not allowed to die, a slave to an idea.

I lay on the hull, my head pressed against the slippery rubber. I thought at times that I belonged to another world but knew not which. A happier world, where no one froze, where salt did not sting the eye. But my hands clung to this world, and for this I thanked God.

Finally, a shy dawn came to my rescue. For the past two hours I had lain on the hull. Hands and legs held fast while my mind wandered. I could no longer control it: dreams and thoughts, reality and hallucination, I could no longer tell the difference . . . a concentration on nothingness . . . but still I stayed alive.

The wind had not lessened its force, but I had to try to right the boat. I could wait no longer. I fixed a long line to the outrigger and pulling from the opposite side, I managed to right her. As the stern pointed westward, I pushed it east. Waves filled the boat. I found to my relief that the compass and my bailing pot had stayed on. The bow pointed far out of the water, the stern just floated. The mizzenmast was broken at its base, the sea anchor lost. I climbed into the boat as waves washed over my face. I looked for three air cushions to push them under the stern. I found the first, inflated it and pushed it far back of the aft washboard. There were the two others, and they followed the first. I began to recover, I drew the spray cover up to my shoulders, untied my pot and bailed. The boat was not yet stable, and I had to sit close to the outrigger.

Big waves ran over the deck, but slowly I put myself in control. I continued bailing. My pot, which held quite a few quarts, struck me as a most useful object. Then I set the course for the west. I bailed until the pot no longer scooped up water, and I finished the job with a sponge.

An hour later the boat was empty, but for some fifty quarts that I left in for ballast. Then I checked the sails, which were in the water on the starboard side. I put them on deck and fastened them. I checked the inside of the boat; every single one of my cans of food had gone, my food supply for the last part of the voyage, had vanished. I had had emergency rations of eleven cans of milk, which I kept in a bag tied to the mast. Where was the flashlight? One had gone, but I found my spare and beamed light at the bow. Clothes, watertight rubber bags lay in chaos. My beam picked out something red. It was the bag full of milk cans. I looked for the two bags that held the Leica, one with black and white film, the other with color. They were gone. The bag with my spare parts was gone. My night glasses, my fluid compass (although I found it later in St. Thomas), all my toilet articles, the grappling iron, all these now floated somewhere in the Atlantic.

The port-side shroud of the mainstay was torn from the deck canvas, the lines for the foresails were in such bad shape that I could not use them without repairing them first.

The frigate bird that had consoled me earlier flew over again. I had to be close to shore. Then I remembered to check my sextant and found it wet in its bag. The chronometer was also full of water but ran the minute I touched it. I took my position before it was too late. The trade winds still blew the storms of yesterday, the storms they might continue to blow for many days to come. The nautical almanac was soaked through; I handled the pages carefully. My latitude was approximately eighteen degrees and twenty minutes, it could not be too far off. My good knife had disappeared. My cans! Gone, too! But I was alive and well, and what more did I want? I did not hoist the sail, as storm and wind pushed me to the west. In the late afternoon I shot a triggerfish with the underwater gun. Its meat tasted better than ever before —and my bottles of rain water had stayed with me.

December 17th

The weather did not change; the sea roared, the storm howled, fish and birds gathered around me again. I was dead tired. During the night my body had shivered uncontrollably. Now at last it was warmer, and the sun shone. Shortly before noon, a wave whistled beside me, reminding me of an old sailor's legend my grandfather had told me when I was a child. The legend of the disaster that a whistling wave brings, springs from the story of a shipwrecked sailor, the sole survivor of a ship that went down shortly after a whistling wave had gone by. The sailors heard it, just as I heard it then. Was the whistling wave an omen for me, too? What did I have to lose? Only my life.

But death did not interest me. Under the water I could see living creatures move; two dolphins. They were small but edible. I shot them, jerked them up on deck, beat them to death with my round knife and devoured them whole. My stomach was still in an unsettled condition, but I ignored its complaints.

Air . . . nothing . . . air at last. I was capsized again! Again I clung to the boat; it was slippery from long thin algae and the few barnacles that were left did not offer much hold. I pushed the boat in the right direction, and there on the bow was a small bulge on the rubber. It was the line from the outrigger, which I had deliberately left there, under the boat. It was easy to right the *Liberia* with the line. Soon I was climbing inside again. I bailed while my elbows held the spray cover high. How had it happened again? My underwater gun had slipped away. I had not had a chance to fasten it after the last dolphin. That was bad. I continued bailing. The islands had to be close by now. Even without the gun I would succeed. I realized that I had capsized over the outrigger each time. Were the Polynesians right when they put it on the windward side? Both times I had felt no shock. It could have been such a sweet death. I had kept my perceptions and my breath. I had to bail. I noticed a bulge in the rubber. It must have been made by something pointed. I decided to try leaving more water in the boat, if the wooden frame could stand it.

I still sat in water. It was not cold; I felt nothing. I had to rig the mizzen sail. It would give the boat speed and lessen the danger of being capsized by angry waves for the third time. With my knife I sharpened the point of the mizzenmast and put the paddle mast on it. It was shaky, but it held. Then I hoisted my sail, relieved to be under sail again and ready to celebrate Christmas then and there.

December 19th

It still stormed; I was empty, a shell, unthinking, kept going only by a complete concentration on the words, "Keep going west, never give up, I will make it." I dozed. Sometimes, for no reason at all, I felt happy; I was somewhere where I could take refuge in irresponsible happiness, where I could escape my ego and my consciousness.

Then I had to find my eleven cans of milk in their red rubber bag near the foremast. Only eleven cans! I felt like drinking two a day. The islands had to be close, I told myself, but these cans were emergency rations and they had to remain that. There was always the possibility that I might capsize again, that I might have trouble with the rudder; I had to save my milk. I would have to drink the water in my bottles; there was still enough to last for a few days. I knew of castaways who had survived more than ten days by drinking only one glass of water a day.

A sudden shock awakened me. The steering mechanism no longer worked, I looked back, drew at the steering cables and—to my horror—found the rudder blade barely hanging on. With a paddle I guided the port-side cable over the stern and pulled the rudder in starboard out of the water. Now

I urgently needed a sea anchor. The rudder wick was broken; small wonder in such weather, but I now faced the problem of a new wick. My spare parts were all gone. Inspiration came to me when I remembered the small wire on top of the mizzen that kept the sail in shape. It was the very thing I needed. With my paddle I kept the boat more or less on course, pulled the wire into the right shape for a wick, slid fully clothed into the water and fastened the rudder blade. My clothes were now completely drenched, but then they had not been fully dry for days. A pathetic afternoon sun occasionally dried me out here and there. At night I invariably sat in water. It was a miracle that my buttocks were not more painful. Was my skin accustomed to the constant immersion?

Clearly before me, in a mind's eye sharpened by danger, I could see the red roofs and green palms of Phillipsburg.

December 21st

I made a sea anchor of my last pieces of string tied to a seabag. But I did not use it, and I hoped I would not have to, for it could only hold me back. A terrible sense of urgency took hold of me, forcing me to calm my nerves with consoling speeches.

"You will make it; keep going west," I repeated endlessly.

When my mind wandered I felt gay and lighthearted, whereas my conscious moments brought tension and worry. There were times when I forgot everything, when I removed myself to a place where there was nothing but an eternal stillness, where I hardly existed and the noise of the storm could not follow.

Frigate birds, circling the *Liberia*, announced my imminent arrival; I knew I was not far from the islands.

My strength lay in my foldboat. Taunting the sea, she

resisted all combers that sought to destroy her. The sea sensed the *Liberia*'s defiance, and the giant waves hounded me for it, doing their best to catch me.

"As long as the mizzen is hoisted, you won't get me," I shouted at them.

Yesterday I drank a can of milk, and today I was greedy for a second. I scanned the sky for clouds that would indicate an island. Nothing! Tomorrow might bring better luck.

During the night I froze; my teeth chattered, my arms shook and I suffered from terrible cramps. "I will make it," I repeated, as I prayed for alert senses. All my life I have prayed; it narrows my consciousness. Now, as I prayed, I concentrated on my arrival in St. Martin, with its colorful houses and fat, green palms.

Why not take a can of milk? I would still have nine left. But what if I had another accident? On a sudden impulse, and with the indifference of an exhausted man, I grabbed for a can, beat with my knife at the edge and sucked out the milk. Only after I had emptied the can, did I have the strength to feel ashamed of my weakness.

December 24th and 25th

The days seemed shorter, the nights longer; nighttime belonged to the devil, while the days belonged to hope. Trade winds blew at thirty to forty miles an hour, bringing with them towering waves. How long had I sailed with no sleep? Many, many days, and God alone knew how many nights. It could have been weeks. The passage of time no longer held meaning for me. Time, Philistine word, a modern, sick word; time, the disease of today!

And today was Christmas Eve. Last year I had discovered my America on this very night. I felt certain it would happen again. "I am lucky," I thought, "and twice I will arrive on Christmas Eve." Royal terns squealed and quarreled around me; my Christmas present, a frigate bird, flew by. I sang Christmas carols.

I wondered which island lay ahead. I could be near Antigua, of course, as I had been last year, but I desperately wanted to arrive in St. Martin. I had set my heart on Phillipsburg.

I checked the sextant; it had rusted and was out of order. Still, I intended to land in Phillipsburg. I was determined. I could not put my hand on my movie camera, but as it was probably useless by now, I decided it didn't matter. I had wanted to take beautiful pictures of waves, to show a storm, even take a picture of a wave running over the boat. What difference did all my lost plans make now?

Christmas Eve. I thought of Christmas trees, of all the great variety of trees I had seen in my lifetime of travel, decorated as Christmas trees. In Liberia, one friend had taken a dead tree, decorated it with seaweed, making, in this fashion, one of the prettiest Christmas trees that I had ever seen. I decided on my Christmas present for that night; a can of evaporated milk. I would drink it at dusk, to remind me of my childhood, when I had always received my presents at twilight on Christmas Eve. I still had seven cans. Should I drink two? After all, I might reach Antigua during the night.

The rudder! Something was wrong. I looked back and saw the rudder was gone, empty hinges stared back at me. The paddle! I had to steer the boat with a paddle, as I had done off the coast of the Sahara the year before. Why, why did this have to happen on Christmas Eve? Better not ask. I did not dare ask. My feet were free for the first time since my departure from Las Palmas. They had accustomed themselves to holding the westward course and were unused to their new liberty. With the paddle my speed decreased considerably. One shoulder felt the brunt of the wind side. The seams of my jacket had burst there, and rheumatic twinges shot through my upper arm, necessitating a change of paddle from starboard to port. It was difficult to keep on course—more so now than ever.

Dusk set in. I punched a small hole in the milk can, nothing came out when I raised it to my lips. The milk had curdled. I banged the can against the metal tip of the paddle and enlarged the hole. The end of my paddle smelled of walnut. How I love walnut cake! Walnut cake with marzipan will always send me back to Germany.

A streak of light glimmered in the sky. The lights of St. John's in Antigua? One voice said yes, urging me to take another can of milk. Another voice urged caution; don't be rash, it counseled. At least, at St. John's I would find English toffee. Then the lights went out, the sky darkened. What a pity, for my sake, that the inhabitants went to bed so early; they could have stayed up a little longer. Tomorrow, I would hit St. John's. Tomorrow—all day long—I would chew on English toffee.

The boat ran backwards. I could not understand it, right there, before Antigua. I was attacked by a dizzy spell; with cramped hands I held the flashlight over the compass. West, it answered me. Then I knew nothing could go wrong. Two flying fish fell into the boat. Should I eat them? No! Not on Christmas Eve, tomorrow I would have all the food I needed in Antigua, so I threw them back into the water. A big comber pushed me forward—then another pulled me back—with both hands I clung to the washboard. The world spun around me. Through my dizziness I could not hear, but knew I was still going back and back. Again two flying fish landed on the boat.

"I warned you," I told them, "not to try this again," and I bit into one. It was full of scales. I scraped them off with my thumb and ate both fish. The sensation of sailing backward was confirmed when I put my hands into the water, although I knew it was an illusion brought on by weariness. My course was west. Again a big comber rushed over the boat, shaking my senses into some kind of reality.

The African had returned.

"Where are we going this time?"

"To the west," he reassured me.

Everywhere I saw the flat shadows of shrubs and bushes. . . . It was flat there, the surf washed over the boat, once, twice . . . oh God! how many times . . . yes! it was so flat there that land had to be somewhere near. My eyes saw what was not there to be seen; my ears heard sounds that did not exist. Suddenly all was quiet. I heard no crash of surf or sound of waves. Why this silence? Why? Why? Only when I talked with my boy or when he answered did I hear anything. How hard he worked. I was happy, endlessly happy, in another world. In a world where the sun shone, where neither body nor spirit existed, a world of ether that surrounded me with irrational joy. I sank even deeper into my mind.

Then I heard again the roaring of the sea, first from the distance, but coming slowly closer. Now it was fresh, menacing and rumbling in my ears. I no longer saw any lights of Antigua. I was awake, alert as never before. My sextant was not working. I concentrated with all my powers on thoughts of Phillipsburg—on the great bay—and the roaring of the sea dimmed. I saw the church, a little church with its roof just over the water and men taking shelter under it. Gray—gray—everything looked gray. Why?

December 26th

I awoke with a dismal sense of oppression. Could I have passed the islands without noticing them? I would not be the first to have missed them. There ought to be land near by. The day before I had seen nine frigate birds and royal terns, a sure sign of land. It had to be somewhere. Should I change course? But how? For a long time now I had not known my exact latitude. I knew I could not be too far to the north, for the trade winds had continually pushed me southward. Still, I decided I would turn south on the first of January. I gave myself five more days to reach St. Martin on the present course.

I still had five cans of milk and a little rain water, brackish but drinkable. Under my skin there still lay a little fat. Oh! I was rich! I could reach the mainland, if necessary. I would make it. Sargasso weed floated on the surface. I caught some, shook it out over my spray cover and feasted off the delicate sea food I found. Several small crabs and two sargasso fish also fell out of the weed, but even in my present situation they had too little flavor to be palatable. However, I enjoyed some little shrimps that I wiped off the plastic cover with my sponge.

Trade winds blew in gusts of twenty-five miles an hour. The easing of the stiff wind made me strong and happy, confident that even without a sextant, I would hit Phillipsburg.

December 28th

There was no longer any doubt that I had missed the islands; frigate birds and terns decreased in number, sargasso weed thinned out. I saw few Madeira petrels or tropic birds. A booby visited me. I tried hard to find out where these birds came from, but I never succeeded. They simply arrived.

December 28th was my birthday. I was hardly aware of it except for passing thoughts of my birthday cake. Cake and Phillipsburg chased each other through my mind.

December 29th

I sat on the *Liberia*, nursing four cans of milk and a hope that never faded. Trade winds had softened to twenty miles an hour or less, the sun shone. It was unbelievable after the long period of stormy seas. Was the sea, at last, becoming reasonable? Soon I expected to sit upon the washboard and enjoy my "hour of hygiene" again. During the day I bailed until my seat was dry once more. But it did not really matter; I was happy—full of hope—and buoyed by the certainty that I would succeed.

Close to noon I noticed a shadow on the horizon to port. A shadow of a cloud, such as I had seen before. No? Yes? I was sure, suddenly, that it was an island. I could see the island, but I could not be sure which one it was. It lay, solitary, on the horizon. I could see it and no other. Strange! I continued sailing through the night, although the island disappeared behind a bank of clouds.

December 30th

I passed a quiet night. I had to sail athwart and to the south to be certain of being in position for the island I had sighted. I was overwhelmed at the thought of seeing land. The island emerged with the sun, bare and broken up by rocks. To starboard lay great, bare rocks; to the north I sighted my actual goal. In the background I could make out another shadow, which I recognized as Saba, the volcanic island. That meant that St. Bartholomew lay ahead of me.

I would land in the bay of Phillipsburg. Now at last I was sure. To the west of St. Martin lies the Anegade Passage, which I would have to cross to reach St. Thomas. It is not easy sailing, and I knew I would have to put my sails and equipment in order, before attempting it.

A last squall stirred the surface of the sea. Slowly I paddled toward St. Martin; rising ahead of me I saw the reality of my dreams: a little church, red roofs and green palms. I looked on peace and calm. It was late afternoon as I entered the harbor, paddling close to the wharf where a crowd was sheltered from the rain squall. I sailed straight through the surf up onto the beach.

Forgotten were my seventy-two days, forgotten my discomforts, my fright and my despair. As I climbed out of the boat, a breaker poured a bucket of water into it, my knees buckled, and I held onto the edge of the boat. Turning her on her bow, I tried to pull the stern out of the water. I went to grab the bow, but I stumbled and fell into the last licks of surf. I tried again, and again I fell, until at last people came over from the wharf and carried the *Liberia* ten yards up the beach to where the water could not reach her. They asked me where I came from. "Las Palmas," I answered, but it meant nothing to them and they ran back to their dry shelter.

I made a stab at clearing up the sails, until I was interrupted by a voice from the pier; a police officer wanted to see my papers, so I drew a watertight bag from under the forepart of the spray cover and stumbled over to him. I handed him my passport and answered his questions. The crowd listened to my story and, after they had grasped the extent of my voyage, insisted on escorting me to the hotel. But the boat was still my first concern. I made my way awkwardly back to her, fastened the sails, closed the spray-cover opening and took out a few of my possessions. Then, very slowly, I walked to the hotel. Questions were thrown at me from all sides, but I hardly heard them. "Seventy-two days at sea, seventy-two days at sea!" repeated itself rhythmically in my mind. I could not believe that I was finally and at last stumbling through the streets of Phillipsburg. Although I walked shakily, like an old man, I did not need support from anyone.

At the hotel, the manager told me that word of my arrival had reached the Governor, who had made me a guest of the island. Someone found clothes for me, and I was led to a shower. Staring back at me, from the old cracked mirror, was a face I did not recognize: sunken eyes, hollow cheeks and unkempt blond beard. Good Lord! Was that my face?

I showered, and then, prepared to face a barrage of questions, I returned to the main room of the hotel, which was crowded with visitors. I sat down to eat and was offered—a cake! a beautiful cocoanut cake! I gulped three slices. As I ate, more and more people crowded into the room, stared at me shyly, shaking their heads in disbelief.

Later that day I investigated my movie camera and found it had corroded, although I hoped to be able to save some of the film. I spread my possessions all over the hotel room to dry before I lay down to sleep. But, still in the grip of immense tension, I could not sleep. At midnight I got up and walked down to the beach, where a small, dark lonely shape lay on the sand. The *Liberia*, a nothing without me. But as I sat down beside her, I thought that without her I, too, would be a nothing. So I sat beside her in unspoken companionship, listening to the surf, whose endless roar calmed me more than the unusual stillness of the hotel bedroom.

The next day a mechanic fixed the rudder, I rigged the boat, stowed away new food supplies and, after another night's rest, left the hospitable island of St. Martin. After fifty hours of comfortable, relaxed sailing, I was before St. Thomas, exactly seventy-six days after I left the Canary Islands. As I paddled into the yacht basin, against the wind, I saw my friends ready to welcome me. They had evidently seen me coming into the harbor. Their first teasing words of greeting reached me as I pulled up alongside the mole. Slowly, with their help, I climbed out of the boat. Someone came over, whispered in my ear, "Didn't you tell me last time that you never intended to do this again?" but as she said it she took my right hand in both of hers, as if to say, I understand you.

8 CONCLUSIONS

I heard later how my friends happened to be gathered at the St. Thomas Yacht basin to see me arrive. I had sent one of them a letter from Las Palmas in which I mentioned that I expected to be with them around Christmas but without saying how I planned to get there. And then one morning a cockleshell sailed from the ocean into the bay of Charlotte Amalie, someone spotted it, and word went around at once, "Here comes Hannes. Have your cameras ready." Thus not only was I warmly greeted on my arrival but I now have in my possession two treasured unposed pictures of myself at that moment.

I spent a few weeks in St. Thomas, regaining my strength and avoiding the present-day Caribbean pirates, who saw a way of making money out of my voyage. For the first few days I had to take antibiotics against abscesses around my knees, but soon they disappeared and I swam and danced again. At the hospital I spent my money in check-ups only

178 Second Voyage

to find that there was nothing extraordinary in my physical condition. My weight had gone down by some forty-three pounds. But as the hospital examination occurred five days after my arrival, I think I probably lost about fifty pounds, most of them during the last few days of the crossing.

I retain fond memories of my interview with the chief of immigration in St. Thomas. I had to go to him to obtain a visa for the United States, which entailed filling out an official questionnaire for Washington. "How would you describe your boat?" he asked, "motorboat, freighter or steamer?"

"Just put down 'other,'" I told him.

"What was your position on the boat? Master, mate or ordinary seaman?"

"Put down 'other,' again."

"What is the boat's tonnage, and how much water does she draw?"

"She weighs fifty-nine pounds, full she draws eight inches, empty about two inches."

Despite my startling answers I was given a visa, and shortly thereafter the *Liberia* and I flew to New York.

I sat in the plane, looking down on the blue canvas of the sea, on which the wind painted white foam ridges. How harmless they looked from above! But I knew what these combers meant. The motors of the plane ran smoothly and rhythmically, passengers read and dozed, a pretty girl came down the aisles, carrying coffee to the pilots. But what, I wondered, would happen to these people if the motors failed, and the plane was forced to ditch?

During my two trips—two hundred days and nights alone at sea—I had learned a great deal that could help castaways. I know now that the mind succumbs before the body, that although lack of sleep, thirst or hunger weaken the body, it is the undisciplined mind that drives the castaway to panic and heedless action. He must learn command of himself and, of course, of his boat, which is often his strongest and most resilient ally. Morale is the single most important factor in survival. Prayer, which brings hope and with hope optimism and relaxation, is a powerful aid in self-mastery. I cannot overemphasize the importance I place on a concentration on strengthening phrases such as those I repeated to myself during the second voyage. Hunger brings on quarrelsomeness, suspicion and irritation in people, so it is well to remember to watch one's neighbor, who may suddenly become the victim of dangerous hallucinations. He may think he sees a food store near the boat and jump overboard to reach it. I had this urge myself several times.

Stimulants are harmful for they are usually followed by a breakdown. Sleep is a vitally important factor, for lack of it leads to delirium, as I know from experience. The castaway should try to sleep, if only for a few minutes at a time. Seconds of sleep may save his life.

Fresh water is another key to survival. Research has been done that proves that a man can survive for two days in the tropics and nine days in temperate zones without food and water. We know that in the last war, a man survived for eleven days in the temperate zone under these conditions. I feel very strongly that no one who wishes to survive should drink salt water. If there is sufficient fresh water on board, a small amount of salt water may be drunk as a salt replacement, but that is all. Salt water is never a substitute for fresh water. If a castaway should happen to have milk or beer on his life raft, he can consider himself fortunate, for both these fluids will give him necessary calories. A skilled fisherman can keep himself supplied with all the solid foods he needs on a life raft, but he must be careful to balance his solid food intake with liquid intake. Only the eyes, blood and spinal liquids of fish supplied me with fluids; to extract liquid from the rest of a fish's body one needs especially built presses.

Above all, my advice to the castaway is never to give up hope; on my second voyage, for example, I met two steamers in regions that are not crossed by shipping lanes.

Many of these thoughts about my own survival crossed my mind as I sat in the plane from St. Thomas to New York, traversing in a few hours a large stretch of the Atlantic Ocean. And why, I wondered, had I felt challenged to cross the Atlantic in a dugout and a foldboat, when I could have done it with ease and safety in a plane or a steamship. What drove me to test my strength of mind and body to the utmost? I realized that no one answer would satisfy me; the urge for adventure, the quest for scientific knowledge—both played a part. I told myself that man has always searched for the new frontier, pushed for further boundaries and that I, as a man, would have to accept that for my answer.

ABOUT THE AUTHOR

Born thirty-five years ago near Hamburg, Germany, DR. HANNES LINDEMANN studied Latin, Greek and French in a Gymnasium there, as well as at Ratzeburg. In 1940 he was drafted into the German infantry and was wounded in Russia. Discharged from the army in December, 1941, he entered the University of Poznan, where he was assistant for the Anatomical Faculty and was active in sports.

In 1943 he was recalled to the Army Medical Corps in Marburg, but continued his medical studies at the University. He later served in hospitals on the Western front. He was sent home in July, 1945, after being captured by the United States Army in Pilsen.

After completing his examinations at the University of Hamburg, he worked for a short while in a Hamburg hospital. Then the desire to travel led him to take a job on a U.S. air base in French Morocco. In 1953 he went to Liberia as plantation doctor for a large American rubber firm, where he worked for two years before his first solo transatlantic voyage.

Dr. Lindemann is planning an around-the-world trip with a larger boat. He hopes that his fiancée will be traveling with him then—as his bride.

NORTH
AMERICA

SOUTH AMERIC

SECOND TRIP
Oct. 1956 - Jan. 1957
FOLDING BOAT

FIRST TRIP
Oct. 1955 - Jan. 1956
DUGOUT CANOE

PORTUGAL
SPAIN
Mazagan
MOROCCO
Safi
SPANISH WEST AFRICA
AFRIC
LIBERIA
GHA

Printed in the USA
CPSIA information can be obtained
at www.ICGtesting.com
LVHW020359221223
767141LV00016B/1115